MATHEMATICS

MATHEMATICS
An Activity Approach

SECOND EDITION

Albert B. Bennett, Jr.
University of New Hampshire

Leonard T. Nelson
Portland State University

Wm. C. Brown Publishers
2460 Kerper Blvd.
Dubuque, Iowa 52001

Copyright © 1985, 1979 by Allyn and Bacon, Inc.

Copyright © 1988 by Wm. C. Brown Publishers. All rights reserved

Library of Congress Catalog Card Number: 85–6014

ISBN 0–697–06852–8

No part of this publication may be reproduced, stored in a retrieval system, or transmitted, in any form or by any means, electronic, mechanical, photocopying, recording, or otherwise, without the prior written permission of the publisher.

Printed in the United States of America by Wm. C. Brown Publishers
2460 Kerper Boulevard, Dubuque, IA 52001

10 9 8 7 6 5 4 3 2

Acknowledgments

Graphic art drawn and supplied by Albert B. Bennett, Jr., and Deborah Schneck.

Unless otherwise stated, the following credits refer to photographs.

CHAPTER 1 CREDITS

pg. 2 (top right) Courtesy of British Information Services. **pg. 5** (middle right) Photo by Ron Bergeron, University of New Hampshire.

CHAPTER 3 CREDITS

pg. 47-48 (cross-number puzzle) Courtesy of Peter Warren, graduate student, University of New Hampshire. **pg. 58** (top right) Courtesy of Stanislaw M. Ulum, University of Colorado. **pg. 60** The name Cuisenaire ® and the color sequence of the rods, squares and cubes, are trademarks of Cuisenaire Company of America, Inc. **pg. 64** (top right) Photo by Ron Bergeron, University of New Hampshire. **pg. 67** Photo by Talbot Lovering.

CHAPTER 4 CREDITS

pg. 75 Courtesy of Museum of Comparative Zoology, Harvard University. **pg. 76** (top right) Courtesy of John P. Adams, University of New Hampshire. **pg. 77** (bottom right) Courtesy of International Business Machines Corporation. **pg. 80** (top right) "Two Intersecting Planes," by M.C. Escher. Courtesy of the Escher Foundation, Haags Gemeentemuseum, The Hague. **pg. 81** "Stars," by M.C. Escher. Courtesy of the Escher Foundation, Haags Gemeentemuseum, The Hague. **pg. 82** Photos by Talbot Lovering.

CHAPTER 5 CREDITS

pg. 102 (top right) Photo by Talbot Lovering.

CHAPTER 6 CREDITS

pg. 105 Fraction Bars from the program, *Fraction Bars*, by Albert B. Bennett, Jr., and Patricia S. Davidson, ©1973. Reprinted by courtesy of Scott Resources, Inc., Fort Collins, Colorado. **pg. 173** Decimal Squares from the program, *Decimal Squares* by Albert B. Bennett, Jr.,©1982. Reprinted by courtesy of Scott Resources, Inc., Fort Collins, Colorado.

CHAPTER 8 CREDITS

pg. 156 (middle) Courtesy of McDonnell Planetarium, St. Louis, Missouri. Photo by Jack Zehrt.
pg. 157, Photos by Ron Bergeron, University of New Hampshire. **pg. 161** (bottom) Reproduced from *Line Designs,* by Dale Seymour and Joyce Snider. Reprinted by permission of Creative Publications, Palo Alto. **pg. 162** (bottom left, middle, and right) Photos by Talbot Lovering. **pg. 164** Photo by Ron Bergeron, University of New Hampshire.

CHAPTER 9 CREDITS

pg. 182 (top quote) Reprinted from *Ford Times,* courtesy of Ford Motor Company. (top right) Photo by Ron Bergeron, University of New Hampshire. (bottom right) Courtesy of the B.F. Goodrich Company. **pg. 183** (top) Courtesy of the B.F. Goodrich Company. **pg. 184** (bottom right) Photo by Ron Bergeron, University of New Hampshire. **p. 186** Photos by Ron Bergeron, University of New Hampshire.

CHAPTER 10 CREDITS

pg. 188 Photo by Dr. B.M. Shaub.

Contents

Preface ix

CHAPTER ONE—MATHEMATICAL REASONING AND PROBLEM SOLVING 1
- 1.1 Geometric Number Patterns 2
  Just for Fun—Fibonacci Numbers in Nature 5
- 1.2 Games of Reasoning 6
  Just for Fun—Game of "Hex" 10
- 1.3 Tower of Brahma 11
  Just for Fun—"Instant Insanity" 13

CHAPTER TWO—SETS, NUMERATION, AND COMPUTERS 15
- 2.1 Sorting and Classifying 16
  Just for Fun—Games with Attribute Pieces 20
- 2.2 Models for Numeration 20
- 2.3 Computer Games in Logo and BASIC 25
  Just for Fun—Mind Reading Cards and Game of "Nim" 28

CHAPTER THREE—WHOLE NUMBERS AND THEIR OPERATIONS 31
- 3.1 Adding with Multibase Pieces and Chip Trading 32
- 3.2 Subtracing with Multibase Pieces 37
- 3.3 Multipying with the Abacus and Chip Trading 41
  Just for Fun—Cross-numbers for Calculators 47
- 3.4 Dividing on the Abacus 48
  Just for Fun—Calculator Games and Number Tricks 52
- 3.5 Patterns on Grids 53
  Just for Fun—Spirolaterals 58
- 3.6 Factors and Multiples with Cuisenaire Rods 60
  Just for Fun—Star Polygons 64

CHAPTER FOUR—GEOMETRIC FIGURES 68
- 4.1 Rectangular and Circular Geoboards 69
  Just for Fun—Tangram Puzzles 74
- 4.2 Regular and Semiregular Tesselations 75
  Just for Fun—Escher-type Tesselations 80
- 4.3 Models for Regular and Semiregular Polyhedra 81
  Just for Fun—Pentrated Tetrahedron 85
- 4.4 Creating Symmetric Figures by Paper Folding 86

CHAPTER FIVE—MEASUREMENT 90
- 5.1 Measuring with Metric Units 91
  Just for Fun—Metric Games 94
- 5.2 Areas on Geoboards 95
  Just for Fun—Pentominoes 98
- 5.3 Models for Volume and Surface Area 99
  Just for Fun—Soma Cubes 103

CHAPTER SIX—FRACTIONS AND INTEGERS 104

- 6.1 Models for Equality and Inequality 105
  Just for Fun—Fraction Games 110
- 6.2 Computing with Fraction Bars 111
  Just for Fun—Fraction Games for Operations 115
- 6.3 Models for Operations with Integers 116
  Just for Fun—Games for Negative Numbers 119

CHAPTER SEVEN—DECIMALS: RATIONAL AND IRRATIONAL NUMBERS 122

- 7.1 Models for Decimals 123
  Just for Fun—Decimal Games 130
- 7.2 Operations with Decimal Squares 131
  Just for Fun—Decimal Games for Operations 136
- 7.3 Computing with the Calculator 137
  Just for Fun—Number Search 142
- 7.4 Irrational Numbers on the Geoboard 143
  Just for Fun—Golden Rectangles 147

CHAPTER EIGHT—GEOMETRY AND ALGEBRA 149

- 8.1 Geometric Models for Algebraic Expressions 150
  Just for Fun—Algebraic Skill Games 155
- 8.2 Conic Sections 156
  Just for Fun—Line Designs and Coordinate Games 161
- 8.3 Geometric Patterns and Number Sequences 164
  Just for Fun—What's My Rule? 167

CHAPTER NINE—MOTIONS IN GEOMETRY 168

- 9.1 Translations, Rotations, and Reflections 169
- 9.2 Devices for Indirect Measurement 174
- 9.3 Topological Entertainment 181

CHAPTER TEN—PROBABILITY AND STATISTICS 187

- 10.1 Probability Experiments 188
  Just for Fun—Buffon's Needle Problem 192
- 10.2 Compound Probability Experiments 193
  Just for Fun—Trick Dice 196
- 10.3 Applications of Statistics 198
- 10.4 Statistical Experiments 201
  Just for Fun—Cryptanalysis 207

Answers to Selected Activities 209

Index 231

Materials Appendix 239

Preface

*Sensitivity to the nature of human learning reveals that learners must have the opportunity to bring some order and structure to mathematics themselves before they are confronted with the more perfect orderliness developed by others.**

Lloyd F. Scott

TO STUDENTS

The side of mathematics that most students see involves computation, definitions, and word problems. In this book you will see another side of mathematics. You will perform experiments, construct models, organize data, look for patterns, and form generalizations from specific instances. You will arrive at and verify your own conclusions, as advocated by Lloyd F. Scott in the above quote, rather than have previously found conclusions presented to you.

There are 10 chapters with from 3 to 6 activity sets in each one. There are 42 pages of materials at the end of this book for use with the activity sets. We encourage you to cut out these materials and become involved in the activities and games. Twenty-eight of the activity sets are followed by *Just for Fun* activities, which are designed for artistic creation, enrichment, and entertainment. The starred (★) activities are either completely or partially answered for your convenience. A complete answer guide may also be purchased.

Many students have found this activity approach to be nonthreatening, enabling them to gain insights into mathematics. As one student said after completing a course, "The activity sets helped me to understand mathematics without memorizing—I have learned to trust my own mind."

TO INSTRUCTORS

This is a book of inductive activities and experiments for learning introductory mathematics. It enables students to experience mathematics intuitively through the use of models and the discovery of patterns. It offers a change, not in content but in pedagogy, to the more structured deductive courses in mathematics.

There are two primary reasons for a book of this nature. First, a substantial amount of mathematics can be taught through inductive and empirical activities. Second, these activities involve the use of concepts, skills, materials, and methods which can be adapted to school mathematics.

*Lloyd F. Scott, "Increasing Mathematics Learning through Improved Instructional Organization," *Learning and the Nature of Mathematics,* William E. Lamon, editor (Chicago, Illinois: Science Research Associates, Inc., 1972), pp. 19–32.

Special Features

Topical coverage—There are activities for most of the topics from number systems, geometry, probability, and statistics which are taught in introductory mathematics courses.

Activities—The questions and activities in each activity set are sequentially developed. There are several activities for each manipulative or model to enable the reader to become familiar with the material and to proceed with depth into the topic.

Materials—There is a Materials Appendix which contains manipulatives, models, game mats, grids, and other materials for the activity sets.

Pedagogy—The activity sets on whole numbers, fractions, integers, and decimals involve models for understanding and teaching the basic operations.

Metric system—There are games and activities for learning the metric system in Chapter 5. Metric units are used in several activity sets following this chapter.

Calculators—Some activity sets have investigations to carry out with a calculator. They are designed to teach mathematical concepts as well as the use of the calculator.

Just for Fun—Twenty-eight of the activity sets are followed by *Just for Fun* activities. These are related to the topics in the activity set.

Organization and Format

There are 36 activity sets under 10 chapter headings. The familiar chapter headings will be convenient for those who wish to use these activities with available mathematics textbooks. The activity sets are mostly independent and may be selected out of sequence. The text, *Mathematics: An Informal Approach,* Second Edition, was written to accompany this activity book. It contains 10 chapters and 36 sections. Each of these sections corresponds to an activity set. Answers are provided for selected activities. These activities are marked with a star. Complete answer guides are available for the activity book as well as the text.

This book can be used in both content and methods courses for the following types of class formats.

A lecture course taught from a text or notes, with the activity sets as supplemental and used as outside assignments.

A combination lecture and lab course in which lectures are followed by a lab with the related activity set.

A lab course based on the activity sets, with outside readings in textbooks and journals.

Audiences

The activity sets are appropriate for several types of audiences. Teachers of all grade levels will find a variety of activities which can be restructured for their classes. The activity sets contain most of the topics which are recommended for preservice elementary school teachers. These topics (and their sections) include: sets (2.1), numeration (2.2), computers (2.3), whole numbers (3.1, 3.2, 3.3, 3.4), factors and multiples (3.5, 3.6), geometric figures (4.1), polyhedra (4.3), symmetry (4.4), metric system (5.1), fractions (6.1, 6.2), integers (6.3), decimals (7.1, 7.2), and algebra (8.1).

Many of the activity sets contain concepts and methods that are not normally found in undergraduate mathematics courses but are appropriate for both junior high and high school teachers. These activities include: number patterns (1.1, 3.5, 8.3), mathematical reasoning (1.2), problem solving (1.3), polygons and polyhedra (4.1, 4.3), metric system (5.1), area and volume (5.2, 5.3), irrational numbers (7.4), coordinate games (8.2), conic sections (8.2), number sequences (8.3), congruence and symmetry (9.1), indirect measurement (9.2), topology (9.3), probability (10.1, 10.2), and statistics (10.3, 10.4).

1

Mathematical Reasoning and Problem Solving

ACTIVITY SET 1.1

GEOMETRIC NUMBER PATTERNS

How long would it take you to find the sum of the whole numbers from 1 to 100?

$$1 + 2 + 3 + 4 + \cdots + 49 + 50 + 51 + \cdots + 97 + 98 + 99 + 100$$

Karl Friedrich Gauss (1777-1855), one of the greatest mathematicians of all time, was asked to compute a sum similar to this when he was ten years old. His schoolmaster had given the problem to Gauss and his classmates, and as was the custom, the first to get the answer would put his or her slate on the teacher's desk. The problem had barely been stated when Gauss placed his slate on the table and said, "There it lies." Can you see how Gauss might have computed this sum?

Karl Friedrich Gauss

A formula for computing this sum will be developed in the following activities by using square and rectangular arrays of dots. These arrays will be used to create several geometric patterns, which in turn will lead to some remarkable number relationships.

1. The following 2 by 2, 3 by 3, and 4 by 4 arrays of dots contain similar patterns. For each of these geometric patterns there is a corresponding number pattern. Sketch a similar pattern on the 5 by 5 array and write the corresponding equation.

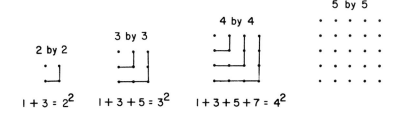

a. Use these patterns to predict the sum of the first ten odd numbers. Check your results by computing the sum.

★ b. These geometric arrays provide strong support for a general statement about sums of odd numbers. Let □ represent the nth odd number, and complete the equation for the sum of the first n odd numbers.

$$1 + 3 + 5 + 7 + \cdots + \square =$$

2. A diagonal pattern on a square array of dots produces a number pattern of increasing and decreasing numbers. Form this pattern on the 5 by 5 array and write the corresponding equation.

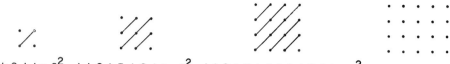

$1 + 2 + 1 = 2^2$ $1 + 2 + 3 + 2 + 1 = 3^2$ $1 + 2 + 3 + 4 + 3 + 2 + 1 = 4^2$

   a. Use the pattern in these equations to compute the following sum:

$$1 + 2 + 3 + 4 + 5 + 6 + 7 + 8 + 9 + 10 + 9 + 8 + 7 + 6 + 5 + 4 + 3 + 2 + 1$$

★   b. Write a formula for the sum of consecutive whole numbers that increase up to n and then decrease to 1.

$$1 + 2 + 3 + \cdots + n + \cdots + 3 + 2 + 1 =$$

3. The diagonal pattern on the rectangular arrays shown next produces a pattern of numbers in which every whole number appears twice. Form this pattern on the 5 by 6 array of dots and write the corresponding equation.

5 by 6

$1 + 2 + 2 + 1 = 2 \times 3$ $1 + 2 + 3 + 3 + 2 + 1 = 3 \times 4$ $1 + 2 + 3 + 4 + 4 + 3 + 2 + 1 = 4 \times 5$

   a. Use the pattern in these equations to determine the following sum. Then check your conclusion by computing the sum.

$$1 + 2 + 3 + 4 + 5 + 6 + 7 + 8 + 9 + 9 + 8 + 7 + 6 + 5 + 4 + 3 + 2 + 1 =$$

★   b. Since each number occurs twice in these equations, there is a more convenient way of writing them. As examples, the equation for the 4 by 5 array can be written as $2 \times (1 + 2 + 3 + 4) = 4 \times 5$, and the equation for the 5 by 6 array can be written as $2 \times (1 + 2 + 3 + 4 + 5) = 5 \times 6$. How can these equations be used to compute $(1 + 2 + 3 + 4)$ and $(1 + 2 + 3 + 4 + 5)$?

★   c. Use the approach in part **b** to compute the following:

$$1 + 2 + 3 + 4 + 5 + 6 + 7 + 8 + 9 =$$

★   d. Use these patterns and observations to compute the sum that Gauss was given.

$$1 + 2 + 3 + 4 + \cdots + 49 + 50 + 51 + \cdots + 97 + 98 + 99 + 100 =$$

★ e. Write a formula for the sum of consecutive whole numbers from one to n.

$$1 + 2 + 3 + 4 + \cdots + n =$$

★ 4. Here are square-shaped patterns on square arrays of dots and their corresponding equations. Form this pattern on the 9 by 9 array of dots and write the corresponding equation.

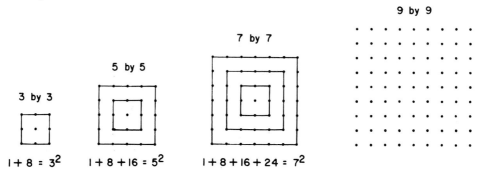

The equation for the 7 by 7 array can be written as $8 + 16 + 24 = 7^2 - 1$. This suggests that one less than the square of an odd number is the sum of consecutive multiples of 8: 8, 16, 24, 32, etc. Try this pattern for the squares of some other odd numbers. For example, one less than 15^2 is 224. Is 224 the sum of consecutive multiples of 8?

★ 5. On the following arrays form square patterns like those in Activity 4. Write the equations that correspond to these patterns.

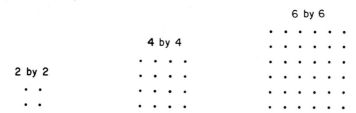

Write similar equations for 8 by 8 and 10 by 10 arrays. Check these equations by computing the sums.

$8^2 = $ _____ $10^2 = $ _____

★ 6. Create your own geometric pattern on the rectangular arrays of dots in part **a** and the square arrays of dots in part **b**. Write the corresponding equations. (One method is to form patterns on the 5 by 7 and 6 by 6 arrays and then draw similar patterns on the smaller arrays.)

★ a.

```
                                              4 by 6           5 by 7
                              3 by 5         · · · · · ·    · · · · · · ·
             2 by 4          · · · · ·       · · · · · ·    · · · · · · ·
            · · · ·          · · · · ·       · · · · · ·    · · · · · · ·
            · · · ·          · · · · ·       · · · · · ·    · · · · · · ·
                             · · · · ·                       · · · · · · ·
```

★ b.

```
                                                       5 by 5        6 by 6
                                        4 by 4        · · · · ·    · · · · · ·
                         3 by 3        · · · ·        · · · · ·    · · · · · ·
           2 by 2        · · ·         · · · ·        · · · · ·    · · · · · ·
           · ·           · · ·         · · · ·        · · · · ·    · · · · · ·
           · ·           · · ·         · · · ·        · · · · ·    · · · · · ·
                                                                    · · · · · ·
```

Just for Fun—Fibonacci Numbers in Nature

The Fibonacci numbers, 1, 1, 2, 3, 5, 8, 13, ..., occur in nature in a variety of unexpected ways. Following the first two numbers of this sequence, each number is obtained by adding the previous two numbers. What are the next five numbers in this sequence?

Daisies: You should have no trouble finding some field daisies, and if you count the petals, there will often be 21, 34, 55, ★ or 89. For which of these numbers will the use of the daisy, to play the game "loves me, loves me not," work to your advantage? The centers of daisies have clockwise and counterclockwise spirals. The numbers of these spirals are also Fibonacci numbers.

Sunflowers: The seeds of the sunflower form two spiral patterns, one proceeding in a clockwise direction and one in a counterclockwise direction. The numbers of spirals in each direction are two consecutive Fibonacci numbers. In this example, there are 34 counterclockwise and 55 clockwise spirals. In larger sunflowers there are spirals of 89 and 144. Find a sunflower and count its spirals.

Cones and Pineapples: Pine, hemlock, and spruce cones have clockwise and counterclockwise spirals of scale-like structures called bracts. The numbers of these spirals are almost always Fibonacci numbers. For example, the white pine cone has 5 and 8 spirals. The sections of a pineapple are also arranged in spirals that satisfy Fibonacci number patterns. Count the spirals on a pineapple from upper left to lower right and from lower left to upper right.

ACTIVITY SET 1.2

GAMES OF REASONING

Inductive reasoning is the process of forming conclusions on the basis of observations and experiments and is like an "educated guess." *Deductive reasoning* is the process of deriving conclusions from given statements. Both types of reasoning are required in the following games. In "Pica-Centro" and "Pick-a-Word" the players have limited amounts of information at the beginning of the game and make conjectures with the aid of inductive reasoning. As more information is obtained, conclusions are based on deductive reasoning. "Patterns" is a game of inductive reasoning. Martin Gardner compares the process of guessing the master pattern in this game to the scientist's search for the laws of Nature. "Players who score high are the brilliant (or sometimes lucky) scientists and poor scorers are the mediocre, impulsive (or sometimes unlucky) scientists who rush poorly confirmed theories into print. Science is like a complicated game in which the universe seems to possess an uncanny kind of order, an order that it is possible for humans to discover in part, but not easily."

1. **"Patterns"* (3 or more players):** Each player draws a 6 by 6 grid of squares and tries to guess the Master Pattern that has been formed by a player called the Designer. The Designer can form any pattern he/she chooses by arranging up to four types of symbols (see Master Patterns 1 and 2) in a 6 by 6 grid. At the end of a round, after all players have had a chance to determine the Master Pattern, another player becomes the Designer.

   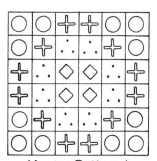

   Master Pattern I

   Play: A player should not see the Master Pattern or the other players' sheets. To obtain information about the Master Pattern, a player decides which squares he/she would like the answers to and puts a small slanted line in the lower corner of each of these squares. The Designer

---

*This game was developed by S. Sackson, *A Gamut of Games* (New York: Random House, 1970) and is described by M. Gardner, "A new pencil and paper game based on inductive reasoning," *Scientific American,* 221 No. 5 (1969), p. 140.

writes the correct symbols in these squares. At this point the player begins to formulate a conjecture about the pattern. To test this conjecture or to get more information, the player can mark more squares with slanted lines and the Designer will fill these in with the correct symbols. Whenever a player thinks he/she knows the Master Pattern, he/she draws the symbols in the remaining squares.

Objective: The object of the game for the players is to guess the Master Pattern by asking for as little information as possible. The object of the game for the Designer is to create a pattern that is fairly easy to discover by at least one player and yet difficult enough to be missed by at least one other player. You will see how these objectives are encouraged by the method of scoring, below. The Master Pattern can take any form, but there should be a strong pattern that the players have a reasonable chance of discovering. In the sample shown in Master Pattern 2 the pattern is 2 by 2 arrays which contain three of the four types of symbols.

Master Pattern 2

Scoring: After all players have their 36 squares filled in, the Designer reveals the Master Pattern. For each square with a correct symbol that was filled in by the player, he/she gets a score of 1 and for every wrong answer a score of ⁻1. The sum of these numbers is the player's score. The player receives no score for the squares filled in by the Designer.

The Designer's score is twice the difference between the best and the worst scores of the players.

Example Grids A, B, and C shown on the next page illustrate the steps involved in arriving at a player's solution. The player put slanted lines in 10 of the squares of Grid A and the Designer drew in the symbols. This information suggests that each column in the Master Pattern has just one type of symbol and that the top half of the grid looks like the lower half. To test these conjectures the player put in three more slanted lines (see Grid B) and the Designer drew in the correct symbols.

Now the player knows that the columns don't all have the same symbol, but it is still possible that the top half of the Master Pattern is the same as the lower half. It also looks as though the outer squares may all have the *three-dot symbol* and that the squares next to them may have only the *circle symbols*.

Using three more slanted lines in Grid C, and getting three more answers from the Designer, seems to verify these conjectures. This appears to be a good point at which to guess the Master Pattern. If all 20 of the remaining squares are filled with the correct symbols, the player's score will be 20.

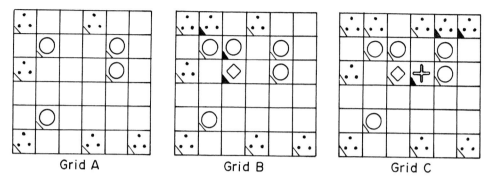

Grid A   Grid B   Grid C

A comparison of the player's solution to the Master Pattern shows that the player is very close. Only 2 of the 20 unknown squares which he/she filled in don't agree with the Master Pattern. Therefore, the player's score is $(18 + {}^-2)$ or 16. Let's suppose the lowest score a player gets for this round is 4. Then the Designer's score is $2 \times (16 - 4)$, or 24.

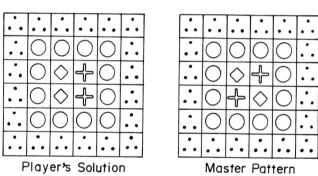

Player's Solution   Master Pattern

2. **"Pica-Centro"*** (2 players): This is a game for two players, although it can be extended to small groups. The objective is to logically deduce a three-digit number that is selected by an opponent.

Preparation: The game begins with Player A choosing a three-digit number, all of whose digits are different. (Zero may be used.) Player B then tries to determine this number by asking questions. Player B keeps a game sheet to record his/her guesses and Player A's responses.

Guesses			Responses	
Digits			Pica	Centro
1	2	3	1	0
3	2	1	0	1
5	4	1	1	1

*See D.B. Aichele, "Pica-Centro, A Game of Logic," *The Arithmetic Teacher,* 19 No. 5 (May 1972), pp. 359-61.

8 CHAP. 1 NATURE OF MATHEMATICS

Play: Player B's first attempt is a guess, but after this he/she begins to use guessing and deductive reasoning based on the information provided by Player A. For example, suppose Player A chooses 471 and Player B's first guess is 123 (see table). Then Player A responds by saying, "You have one pica," which means that Player B has one of the digits in 471 but that it is in the wrong position. A logical second guess for Player B is 321. In this case Player A says, "You have one centro," which means that there is one correct digit and that it is in the right position. Player B can now deduce that either the 1 or the 3 is in the correct position and that 2 is not in Player A's number. How? Player B's third guess is 541, which has one digit in the correct position (the 1) and one digit of Player A's number which is in the wrong position (the 4). Player A's response is, "One pica and one centro," which Player B writes on his/her game sheet. Play continues in this manner until Player B has deduced Player A's number.

Ending: After Player B has guessed Player A's number, he/she will have a 0 in the pica column and a 3 in the centro column. The roles are then reversed with Player B choosing a number and Player A asking the questions. The player who requires the fewest number of turns to guess the opponent's number is the winner.

Variations: Four-digit, five-digit, or numbers with any number of digits can be used as long as the players agree on the number of digits before playing.

3. **"Pick-a-Word" (2 players):** This game is played like "Pica-Centro" except that Player 1 chooses a word, the length of which is agreed upon before play begins, and Player 2 tries to guess what it is. The game sheet shown here is for a four-letter word. Player 1 must choose a word that is in the standard dictionaries, but Player 2 (the player guessing) can guess any four letters in any order.

Guesses				Responses	
Letters				Pica	Centro
a	b	c	d	2	0

If, for example, Player 1 chooses the word "barn," and Player 2's first guess was a, b, c, d, then Player 2 would reply, "You have two pica."

When Player 1 gets the correct word, barn, there will be a 0 in the pica column and a 4 in the centro column.

Just for Fun – Game of "Hex"

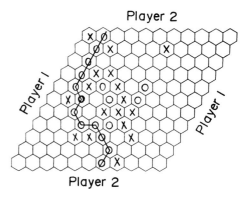

"Hex" is a two-person game with lots of opportunities for deductive reasoning. It is played on a gameboard of hexagons similar to the 11 by 11 grid shown here (see Material Card 1). Player 1 and Player 2 take turns placing their markers or symbols on any unoccupied hexagon. Player 1 attempts to form an unbroken chain or string of symbols from the left side of the grid to the right, and Player 2 tries to form a chain of symbols from the top of the grid to the bottom. The four corner hexagons can be used by either player. The first player to complete a chain is the winner. In this sample game, Player 2 (using O's) won against Player 1 (using X's).

Winning Strategy: There will always be a winner in the game of "Hex" since the only way one player can prevent the other from winning is to form an unbroken chain of markers from one side of the grid to the other. On a 3 by 3 gameboard the person who
★ plays first can easily win in three moves. What should be the first move? Find a first move on the 4 by 4 and 5 by 5 gameboards that will insure a win regardless of the opponent's moves.

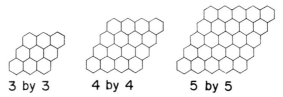

3 by 3   4 by 4   5 by 5

It is tempting to conclude that a winning strategy on an 11 by 11 gameboard begins with a first move on the center hexagon. However, there are so many possibilities for plays on a grid of this size that no winning strategy has yet been found. There is a type of proof in mathematics, called an *existence proof,* which shows that something exists even though no one is able to find it. It has been proved that a winning strategy does exist for the first player in a game of "Hex" on an 11 by 11 grid, but just what that strategy is, no one knows.*

*For an elementary version of this proof, see M. Gardner, "Mathematical Games," *Scientific American,* 197 No. 1 (1957), p. 145.

ACTIVITY SET 1.3

TOWER OF BRAHMA

There is a legend that in the great temple at Benares, India, beneath the dome that marks the center of the world, there rests a brass plate in which are fixed three diamond needles. On one of these needles, during the creation, God placed 64 discs of pure gold, the largest resting on the brass plate, and the others becoming

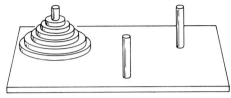

Tower of Brahma

progressively smaller up to the top one. This is the Tower of Brahma. Day and night, unceasingly, the priests would transfer the discs from one diamond needle to another according to the fixed and immutable laws of Brahma, which require that the priest must not move more than 1 disc at a time and that this disc must be placed on a needle so that there is no smaller disc below it. When the 64 discs have been transferred from the needle on which God placed them to one of the other needles, tower, temple, and Brahmans alike will crumble into dust, and with a thunderclap the world will vanish.*

The following activities involve finding the minimum numbers of moves for transferring given sets of discs from one needle to another. For small numbers of discs this is done by experiments. For larger numbers of discs this involves several problem-solving techniques: Drawing pictures; forming tables; looking for patterns; and reducing a problem to a simpler one that has been solved.

★ 1. Use the Tower of Brahma and discs (Material Cards 2 and 3) to transfer 2 discs from one peg to another in the fewest number of moves. How many moves does it take? Repeat this activity for three discs and then again for four discs. Write your results in the table. Then try to predict the number of moves for transferring 5 discs.

Number of Discs	Number of Moves
1	1
2	
3	
4	
5	

2. It takes 3 moves to transfer 2 discs from one peg to another. This fact is used twice in the next sequence of figures. Use these figures to explain why 7 moves are required to transfer 3 discs from one peg to another.

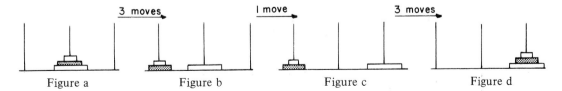

Figure a   Figure b   Figure c   Figure d

*E. Kasner and J. Newman, *Mathematics and Imagination* (New York: Simon and Schuster, 1940), pp. 170–71.

3. The following figures show 4 discs being transferred from one peg to another. Write the number of moves above the arrows for transferring the discs from one figure to the next. How many moves in all will it take to transfer the 4 discs from one peg to another?

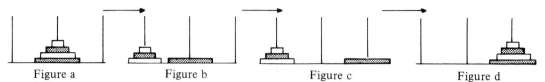

Figure a   Figure b   Figure c   Figure d

4. Use the approach from Activities 2 and 3 to determine the number of moves for 5 discs.

   a. Complete this table for moving up to 12 discs.

Number of Discs	1	2	3	4	5	6	7	8	9	10	11	12
Number of Moves	1	3	7	15								

★ b. Explain how the number of moves for 19 discs can be used to find the number of moves for 20 discs.

5. The Brahman priests must transfer 64 discs. In order to compute the number of moves this will take, notice that each number of moves for transferring a given number of discs is 1 less than a power of 2. Write the number of moves for 4, 5, and 6 discs using powers of 2.

Number of Discs	Number of Moves
1	$2 - 1 = 1$
2	$2^2 - 1 = 3$
3	$2^3 - 1 = 7$
4	
5	
6	
⋮	
64	

★ a. How many moves will it take to fulfill the Brahman prophecy? (Leave your answer in exponential form.)

   b. Approximately how many years will it take to transfer all 64 discs if each move takes 1 second?

6. **The 4-Peg Tower of Brahma:** Suppose that instead of 3 pegs for the tower of Brahma there are 4 pegs. As before, the discs must be moved one at a time and in such a way that no disc is placed on a smaller one.

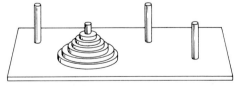

4-Peg Tower of Brahma

★ a. The number of moves for 1 to 11 discs are given in the following table. Find the pattern in these numbers and predict the number of moves for 12, 13, 14, and 15 discs.

Number of Discs	1	2	3	4	5	6	7	8	9	10	11	12	13	14	15
Number of Moves	1	3	5	9	13	17	25	33	41	49	65				

b. The number of moves for 12 discs can be found by the following sequence of moves: Use 4 pegs to move the top 8 discs (33 moves); use 3 pegs to move the bottom 4 discs (15 moves); and then use 4 pegs to move the top 8 discs onto the bottom 4 discs (33 moves). Check your answer in the table for 15 discs by a similar procedure: First move the top 10 discs; then the bottom 5 discs; then the top 10 discs. How many moves will be needed for 16 discs?

7. **Challenge:** If the priests in charge of transferring discs on the Tower of Brahma moved 1 disc every second, it would take them over 500 billion years to transfer all 64 discs. At the same rate, how long would it take them to transfer 64 discs if there were 4 diamond needles instead of 3 needles?

Just for Fun—"Instant Insanity"

"Instant Insanity" is a popular puzzle that was produced by Parker Brothers several years ago. The puzzle consists of four cubes. The faces of each cube are colored either red, white, blue, or green (see patterns in the next diagram). The object of the puzzle is to stack the cubes so that each side of the stack (or column) has each of the four colors. Make a set of these cubes and try the puzzle.

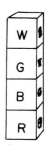

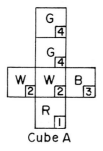
Cube A

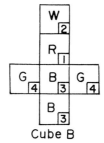
Cube B

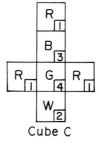
Cube C

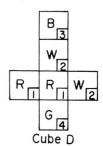
Cube D

The trial and error approach is fun at first, but it soon becomes tedious. Robert E. Levin described the following deductive method of solving this puzzle in the July 1969 issue of the *Journal of Recreational Mathematics.** He numbered the faces of the cubes

*R. Levin, "Solving Instant Insanity," *Journal of Recreational Mathematics,* 2 (1969), pp. 189-92.

(see previous diagram) using 1 for red, 2 for white, 3 for blue, and 4 for green. To solve the puzzle the numbers 1, 2, 3, and 4 must occur down each side of the stack. The sum of these numbers is 10, and for a solution the sum of the numbers on opposite sides of the stack must be 20.

Levin made the following table showing opposite pairs of numbers on each cube and their sums. For example, 4 and 2 are on opposite faces of cube A, and their sum is 6. The bottom row of the table contains the sums of opposite faces. For one combination of of numbers whose sum is 20, the sums have been circled. Notice that above these circled numbers, the numbers 1, 2, 3, and 4 each occur only twice. This tells you how to stack the cubes so that two of the opposite sides of the stack will have a total sum of 20. (Both of these sides will have all 4 colors.) The two remaining sides of the stack must also have a sum of 20. Find four more numbers from the bottom row (one for each cube) whose sum is 20, and such that each of the numbers 1, 2, 3, and 4 from the faces occurs exactly twice. The remaining two faces of each cube will be its top and bottom faces when it is placed in the stack.

	Cube A			Cube B			Cube C			Cube D		
Pairs of Opposite Faces	4	4	3	4	1	2	1	1	3	1	1	4
	2	1	2	4	3	3	1	4	2	3	2	2
Sums of Pairs	⑥	5	5	8	4	⑤	2	⑤	5	④	3	6

14   CHAP. 1   NATURE OF MATHEMATICS

2

Sets, Numeration, and Computers

ACTIVITY SET 2.1

SORTING AND CLASSIFYING

Since their introduction in the late 1960s, sets of geometric figures called attribute blocks or attribute pieces have become a popular means of illustrating relationships between sets and operations with sets.* The attribute pieces on Material Card 25 vary in color, shape, and size. There are three colors (red, blue, and yellow); five shapes (square, hexagonal, triangular, rectangular, and circular); and two sizes (large and small). Each attribute piece is labelled with its attributes: *LRS* is "large red square," *SBC* is "small blue circle," and so on, with the letters of each label denoting size, color, and shape, in that order. These attribute pieces should be cut out. They will be used in the following activities to illustrate disjoint sets, equivalent sets, union, intersection, subsets, and complements. The four attribute pieces shown above are arranged in a row so that each piece differs by exactly two attributes (color, shape, or size) from the piece next to it. Conditions similar to this are used in the "Grid Game," following this activity set. (*Note:* Squares are not to be thought of as rectangles in these activities.)

1. Subsets of attribute pieces can be formed by combining two attributes. The attribute pieces shown here are all small and red. (This is the intersection of the set of small attribute pieces with the set of red attribute pieces.)

   $$SR = \{SRH, SRT, SRR, SRC, SRS\}$$

   Use your attribute pieces to list the elements in the following sets. *BH* is the set of blue hexagons, *LR* is the set of large rectangles, and *SS* is the set of small squares.

   $BH = \{$_____$\}$ $LR = \{$_____$\}$ $SS = \{$_____$\}$

   Use the sets *SR, BH, LR,* and *SS* to answer the following questions.

   ★ a. Which pairs of sets are disjoint (have no elements in common)?

   ★ b. Which pairs of sets are equivalent (can be put into one-to-one correspondence)?

   ★ c. Which sets do the following attribute pieces belong to ($\in$, is an element of)?

   $LYR \in$ _____ $SBS \in$ _____ $SBH \in$ _____

*Z.P. Dienes and E. W. Golding, *Learning Logic, Logical Games* (New York: Herder and Herder, 1970), available from McGraw-Hill; and Elementary Science Study, *Teacher's Guide for Attribute Games and Problems* (St. Louis: Webster Division of McGraw-Hill, 1968).

2. The attribute pieces shown here satisfy
the condition large or rectangular.
(These attribute pieces are in the union
of the set of large attribute pieces and
the set of rectangular attribute pieces.) Notice that these pieces may satisfy both conditions, but only one condition is necessary.

Use your attribute pieces to finish listing the elements in the following sets.

$A = \{SRR, LBH, LYT, LRR$ _____ $\}$

  A contains all attribute pieces that are large or rectangular.

$B = \{LRH, SRS$ _____ $\}$

  B contains all attribute pieces that are hexagonal or red.

★ a. Are sets A and B disjoint?

★ b. Are sets A and B equivalent?

3. Each circle in the accompanying Venn diagram is labelled with a different attribute. All attribute pieces that are red should be placed in set E; all those which are small belong to set F; and set G is for triangular attribute pieces. The small red triangle has all three of these attributes and so it is placed in $E \cap F \cap G$. The attribute pieces that are small and red should be placed in $E \cap F$. Draw large circles for sets E, F, and G and sort your attribute pieces into the appropriate regions.

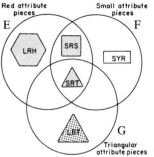

List the attribute pieces in the following four sets.

★ a. $F \cap G = \{$ _____ $\}$
   b. $E \cup F = \{$ _____ $\}$
★ c. $E \cup (F \cap G) = \{$ _____ $\}$
   d. $G \cap (E \cup F) = \{$ _____ $\}$

★ 4. We have been using the attributes of size, color, and shape. Here are some other attributes: all angles congruent; all sides congruent; opposite sides are parallel; all angles equal to 90 degrees.

★ a. Which one of the attribute pieces shown here has all four of these attributes?

★ b. Describe the different types of attribute pieces which are contained in sets B, C, and D. For example, see set A.

  Set A contains an attribute piece if its sides are all congruent (have the same length). All square, hexagonal, and triangular attribute pieces.

ACTIVITY SET 2.1   17

Set B contains an attribute piece if its opposite sides are parallel.
Set C contains an attribute piece if all its angles are right angles (90-degree angles).
Set D contains an attribute piece if its angles are less than 90 degrees.

★ c. Use the preceding sets to determine which of the following statements are true and which are false.

$$B \subset A \qquad C \subset B \qquad D \subset A \qquad C \subset A \qquad A \subset B$$

5. Here are some of the attribute pieces that are not(small and red). One way to find all such pieces is to remove those attribute pieces which are *small and red*  and see what is left. This set of leftover attribute pieces has no elements in common with the set of attribute pieces that are small and red. The union of these two sets is the complete set of 30 attribute pieces. Two such sets having no elements in common but whose union is the whole set are called *complements* of each other.

   a. Finish listing the attribute pieces that are not(small and red).

   $A = \{ LRH, LBR, SYT, SYS,$ _____ $\}$

   b. The two attribute pieces shown here are not small and not red. These pieces satisfy both of the conditions of being not small and not red. Finish listing the attribute pieces that are not small and not red.

   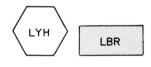

   $B = \{ LBR, LYH,$ _____ $\}$

★ c. Does $A = B$?
★ d. What can we conclude about the condition not(small and red) as compared to the condition not small and not red?

6. It is possible for differently worded descriptions to result in equal sets. There are four descriptions for the following sets, and two of these sets are equal to two of the others. Use your attribute pieces to form these sets. List their elements.

   Set D contains all attribute pieces that are not(small and rectangular). (*Hint:* Find the set of attribute pieces that are small and rectangular and then take its complement.)

   $D = \{$ _____ $\}$

   Set E contains all attribute pieces that are not small and not rectangular. (*Hint:* List all attribute pieces that are not small and then list all those that are not rectangular. Take the intersection of these two sets.)

   $E = \{$ _____ $\}$

Set F contains all attribute pieces that are not(small or rectangular). (*Hint:* To find this set, find the set of attribute pieces that are small or rectangular, and take its complement.)

$$F = \{\underline{\hspace{10cm}}\}$$

Set G contains all attribute pieces that are not small or not rectangular. (*Hint:* List all attribute pieces that are not small. Then list those which are not rectangular, and take the union of the two sets.)

$$G = \{\underline{\hspace{10cm}}\}$$

★ a. Which pairs of these sets are equal?

★ b. The equalities and descriptions of the preceding four sets suggest generalizations about statements involving "not," "and," and "or." For any two attributes A and B, two of the following four statements have the same meaning as the other two statements. Which pairs of statements have the same meaning?

  not(A and B)   not(A or B)   not A or not B not A and not B

★ 7. Find the attribute pieces that satisfy the descriptions in parts **a, b,** and **c.** One of these sets is the empty set. List the elements of the other two sets. (*Hint:* Use your attribute pieces and select those which satisfy the first condition. Then use this set to select those which satisfy the second condition, etc.)

★ a. Set X contains all attribute pieces which are not circular, have 90° angles, are not red, have all sides congruent, and are not large.

$$X = \{\underline{\hspace{10cm}}\}$$

★ b. Set Y contains all attribute pieces which are not(square or large), are blue, have 90° angles, and have all sides congruent.

$$Y = \{\underline{\hspace{10cm}}\}$$

★ c. Set Z contains all attribute pieces which are large or red, have opposite sides parallel, are not red or small, do not have 90° angles, and are not blue.

$$Z = \{\underline{\hspace{10cm}}\}$$

Just for Fun—Games with Attribute Pieces

"Grid Game" (2 or more players or teams): This game begins by placing any attribute piece on the center square of the grid (see Material Card 5). A large red hexagonal piece was used to start the game on the grid shown here. The players take turns placing an attribute piece on the grid according to the following conditions: The piece that is played must be placed on a square that is adjacent, by row, column, or diagonal, to a piece that has already been played; the adjacent attribute pieces in the rows must differ by one attribute (color, size, or shape); the adjacent pieces in the columns must differ by two attributes; and the adjacent pieces in the diagonals must differ by three attributes. A player's score on each turn is the total number of attributes between the piece he/she played and all adjacent attribute pieces.

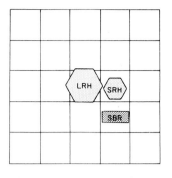

Examples The first piece played was a small blue rectangle, which differs by 3 attributes from the large red hexagonal piece. This player scored 3 points. The second player played the small red hexagonal piece. This differs from the small blue rectangle by 2 attributes and from the large red hexagon by 1 attribute. This player also scored 3 points. The game ends when no more pieces can be played on the grid.

Attribute Game (2-4 players or teams): Draw 3 large circles and label them *X, Y,* and *Z*. Each circle represents a set of attribute pieces. One team (player) decides what the sets are to be, but they do not tell their opponent. The descriptions of the sets should be written down. Then 4 attribute pieces are placed in the circles so that there is at least one in each circle. The opposing team then tries to guess what the sets are by asking yes or no type questions. For example, is *Y* the set of pieces with 90-degree angles? After one team guesses the 3 sets, they reverse roles and play again. The team (player) requiring the least number of questions wins the game.*

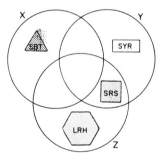

ACTIVITY SET 2.2

MODELS FOR NUMERATION

Multibase Pieces (Activities 1-4): Place value is an important concept for understanding counting and computation with whole numbers. Multibase pieces, which were designed by Zoltan Dienes, form one

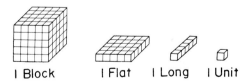

1 Block   1 Flat   1 Long   1 Unit

---

*For a description of four more types of games with attribute pieces, see D.E. Cruikshank, "Sorting, Classifying, and Logic," *The Arithmetic Teacher,* **21** No. 7 (November 1974), pp. 588-98.

20   CHAP. 2 COUNTING, SETS, AND NUMERATION

of several models for illustrating place value. Dienes used different sets of pieces for the bases three, four, five, six, and ten. The pieces shown on the preceding page and below are for base five. Five units equal 1 long; 5 longs equal 1 flat; and 5 flats equal 1 block. There are multiple copies of these pieces on Material Card 31, which should be cut out for use with the following activities.

1. In the following set of multibase pieces, five of the units can be replaced by 1 long, and five of the longs can be replaced by 1 flat. This process is called regrouping. After regrouping as much as possible, this set can be represented by 1 block, 2 flats, 2 longs, and 3 units.

Regroup each set of base five pieces so that there is a minimum number of pieces. After regrouping, write the numbers of blocks, flats, longs, and units in the accompanying tables. Why is it unnecessary to write a number greater than 4 in any of the boxes?

★ a.

b.

2. This next set of multibase pieces has a total of 339 units: There are 250 units in the 2 blocks; 75 units in the 3 flats; and 10 units in the 2 longs. The numbers of the various pieces are shown in the table on the right.

B	F	L	U
2	3	2	4

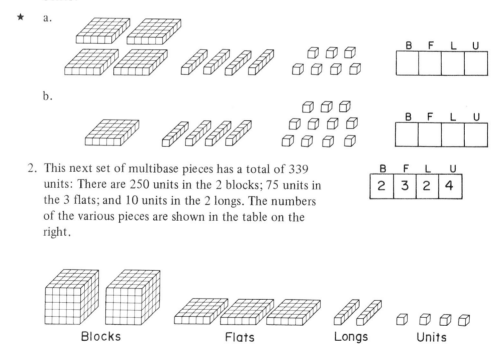

ACTIVITY SET 2.2   21

Use the minimum number of your base five multibase pieces to represent each of the following numbers of units. Record the numbers of these pieces in the given tables.

★ a. 89 units   b. 125 units   ★ c. 137 units   d. 25 units   ★ e. 453 units

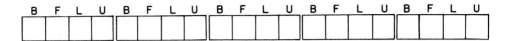

3. The numbers of base five pieces can be recorded without using the tables. For example,

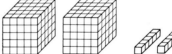

can be replaced by 2034_{five}

In order to do this we must agree that the positions of the digits from right to left represent the numbers of units, longs, flats, and blocks, respectively. This method of writing numbers is called *positional numeration,* and 2034_{five} is called a *base five numeral.*

The base five pieces for the numeral 2034_{five} are shown next. There is a total of 269 units in these pieces. Therefore, 2034_{five} and 269 represent the same number.

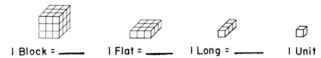

Use your base five pieces to represent each of the following numbers. Then determine the total number of units in each set of multibase pieces.

★ a. 2304_{five} b. 1032_{five} ★ c. 3004_{five}

4. Here are the base three multibase pieces. Write the number of units contained in each piece.

I Block = _____   I Flat = _____   I Long = _____   I Unit

★ a. Use the minimum numbers of base three pieces to represent 70 units. Write the numbers of these pieces in the given table.

B	F	L	U

★ b. Why is it unnecessary to write a number greater than 2 in the boxes of this table?

Abacus (Activities 5–8): The early abacus was a tray covered with dust or sand on which figures could be drawn. Another form of abacus used by the ancient Romans was a ruled table upon which counters were placed. The abacus we will use has markers that can be moved on and off columns. It is similar to the tabletop model used by the Romans. This abacus is on Material Card 7 and the markers are on Material Card 30. With this type of abacus any number of markers can be placed on the columns, enabling us to select different values of b for the base. For example, if we let $b = 5$, the markers on the preceding abacus represent the base five number, 234_{five}. At first you may wish to think of these markers as representing units, longs, flats, and blocks. The 2, 3, and 4 markers on the columns of this abacus represent 2 flats (2×5^2), 3 longs (3×5), and 4 units, respectively.

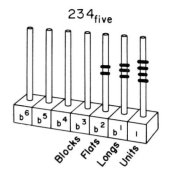

5. Draw markers on these abacuses to represent the given numbers.

   ★ a. 3023_{five} b. 1402_{five}

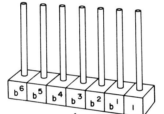 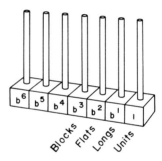

6. For $b = 5$, represent 1204_{five} on your variable base abacus. Each marker on the columns represents a power of 5. For example, the marker on the b^3 column represents 1 block or 5^3. Since 1 block equals 5 flats, the marker on the b^3 column can be replaced by 5 markers on the b^2 column. This is called *regrouping*.

   ★ a. How many markers should be placed on the b column to replace 1 marker from the b^2 column. Why?

   b. How many markers should be placed on the units column to replace 1 marker from the b column? Why?

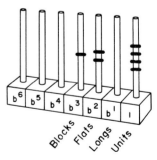

★ c. Use your abacus to regroup the markers that represent 1204_{five}, so that there are 4 markers on the b^2 column and 13 markers on the b column. How many markers will be needed on the units column? Show your results on this abacus.

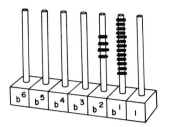

7. Let $b = 7$ and represent 1204_{seven} on your variable base abacus. Each marker on the columns now represents a power of 7.

★ a. How many markers should be placed on the b^2 column to replace 1 marker from the b^3 column?

b. How many markers should be placed on the b column to replace 1 marker from the b^2 column?

★ c. Regroup the markers for 1204_{seven} so that there are 7 markers on the b^2 column and 18 markers on the units column. How many markers will there be on the b column? Sketch the markers for each of these columns on the second abacus.

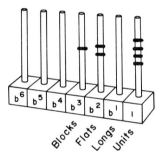 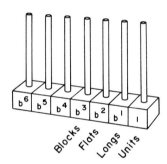

8. With base five, if there are 5 or more markers on a column they can be regrouped. The 5 markers on the b^2 column of the following abacus can be replaced by 1 marker on the b^3 column, since 5 flats equal 1 block ($5 \times 5^2 = 5^3$). Regroup the markers on the first abacus so that there are less than 5 markers on each column. Show your results on the second abacus.

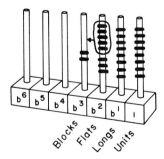 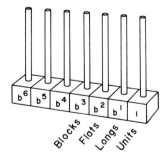

9. Using your variable base abacus, place 8 markers on each of the columns b^2, b, and units. For each of the following bases do as much regrouping as possible and then write the number represented by the markers.

   ★   a. Base five ($b = 5$)

       b. Base seven ($b = 7$)

   ★   c. Base three ($b = 3$)

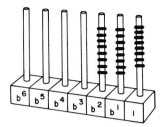

Network III, a computer-generated environment, Walker Art Center, Minneapolis

ACTIVITY SET 2.3

COMPUTER GAMES IN LOGO AND BASIC

The network of lights in the photograph is activated by a computer as people walk on the carpet. The carpet is subdivided into small squares, each containing a pressure-actuated switch that detects the presence of a person on that square. The computer is programmed to determine where people are within the space, which direction they are moving in, how fast they are traveling, etc. The computer uses this information to generate patterns in the network of lights. The sense of a highly elaborate abstract game is intended to be conveyed, tempting the person into testing the environment to see what effect he/she has on the responses.

1. **Target Game (Logo):** A target which is a circle of radius 9 and the turtle are randomly placed on the screen. The objective is to get the turtle inside the circle in as few tries as possible.

   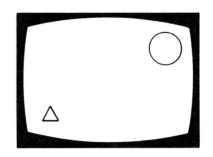

   The procedure TARGETGAME sets up the screen for this game. TARGETGAME is defined by using two other procedures: CIRCLE9 draws a circle of radius 9; and PLACERANDOMLY places the turtle and the circle in random locations.

   a. Type the commands for CIRCLE9, PLACERANDOMLY, and TARGETGAME as they are shown here.

```
TO CIRCLE9                          TO PLACERANDOMLY
  REPEAT 360 [FD .0174 * 9 RT 1]      PENUP
END                                   RT 90 FD ( RANDOM 260 ) - 130
                                      LT 90 FD ( RANDOM 220 ) - 110
                                      PENDOWN
TO TARGETGAME                       END
  PLACERANDOMLY CIRCLE9
  PENUP HOME
  PLACERANDOMLY
END
```

   b. Play this game. Type TARGETGAME to set the game up. A move consists of giving the turtle a direction and or a distance. For example, RT 60 FD 150 is one move. The object of the game is to get the turtle *completely* inside the circle in as few moves as possible. When you have succeeded, your opponent types TARGETGAME and attempts to finish the game in fewer moves.

   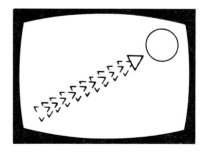

   ★ c. The definition of PLACERANDOMLY uses the command RANDOM, which randomly selects whole numbers from 1 to a given number. For example, RANDOM 360 selects a number from 1 up to and including 359, and RANDOM 100 selects a number from 1 up to and including 99. Explain why the numbers 260, -130, 220, and -110 were used in the definition of PLACERANDOMLY.

d. Revise the game so that the target is a square, with sides of length 18.

2. **Hi-Lo (BASIC):** In this game the computer selects a random whole number less than 1000 for you to guess. After each guess, the computer tells you the number of your guess and whether the guess is too low, too high, or correct. When you have guessed the correct number, the computer prints the total number of guesses.

a. Type the program for this game as defined here.

```
10    LET B =   INT (1000 *   RND (1))
20    PRINT "THE COMPUTER HAS SELECTED A WHOLE NUMBER < 1000.
      TRY TO GUESS THE NUMBER. EACH TIME YOU GUESS THE COMPUT
      ER WILL PRINT THE NUMBER OF YOUR GUESS AND WHETHER YOUR
       GUESS IS TOO LOW, TOO HIGH, OR CORRECT."
30    LET X = X + 1
40    PRINT "PLEASE TYPE GUESS NUMBER "X"."
50    INPUT K
60    IF K < B THEN   GOTO 100
70    IF K > B THEN   GOTO 120
80    PRINT "THAT'S RIGHT! YOU GUESSED THE CORRECT NUMBER IN "
      X" GUESSES."
90    GOTO 140
100    PRINT K" IS TOO LOW."
110    GOTO 130
120    PRINT K" IS TOO HIGH."
130    GOTO 30
140    END
```

b. Play this game a few times and try to minimize your number of guesses. (Begin the game by typing RUN.)

★ c. How can the program be changed so that the computer selects a random number less than 128 (rather than 1000)? If this change is made, what is the minimum number of guesses necessary to identify any randomly selected number?

d. Determine the minimum number of guesses necessary to identify any randomly selected number less than 1000.

3. **Pica-Centro (BASIC):** In this game the computer will select a whole number greater than 99 and less than 1000, all of whose digits are different. You will try to guess this number in as few turns as possible. If you guess 1 digit in the correct place, you get 1 Centro ($C = 1$); if you guess 2 digits in the correct places, you get 2 Centro ($C = 2$); if you guess 1 of the digits, but it is in the wrong place, you get 1 Pica ($P = 1$); if you guess 2 of the digits, but they are in the wrong places, you get 2 Pica ($P = 2$); if you guess 1 digit which is in the correct place and 1 which is in the wrong place, you get $C = 1$ and $P = 1$; etc. The game ends when you have 3 Centro ($C = 3$).

```
10   PRINT "THE COMPUTER WILL SELECT A 3-DIGIT NUMBER (WITH NO TWO
     DIGITS EQUAL) FOR YOU TO GUESS."
15   LET W =  INT (9 *  RND (1) + 1): LET Y =  INT (10 *  RND (1)):
     LET Z =  INT (10 *  RND (1))
20   IF W = Y OR W = Z OR Y = Z THEN   GOTO 15
30   LET K = 0
40   LET C = 0: LET P = 0: LET K = K + 1
50   PRINT "WHAT IS GUESS NUMBER "K" ? TYPE IN YOUR THREE DIGITS SE
     PARATED BY COMMAS."
60   INPUT D,E,F
70   IF D = W THEN C = C + 1
80   IF D = Y THEN P = P + 1
90   IF D = Z THEN P = P + 1
100    IF E = W THEN P = P + 1
110    IF E = Y THEN C = C + 1
120    IF E = Z THEN P = P + 1
130    IF F = W THEN P = P + 1
140    IF F = Y THEN P = P + 1
150    IF F = Z THEN C = C + 1
160  PRINT "YOU HAVE "C" CENTRO AND "P" PICA."
170  IF C = 3 THEN   GOTO 190
180  GOTO 40
190  PRINT "YOU HAVE FOUND THE RIGHT NUMBER, "100 * W + 10 * Y + Z
     ", IN "K" GUESSES."
200  END
```

    a. Type this program and play the game a few times.

★ b. In line 15 three random digits are assigned to W, Y, and Z. What is the purpose of line 20?

    c. How can the program be changed so that repeated digits are allowed for the 3-digit number which is selected by the computer?

Just for Fun: **Mind Reading Cards and Game of "Nim"** (Mind Reading Cards–Material Card 8).

The numbers 1, 2, 4, 8, 16, 32, etc., are called *binary numbers*. Following 1, each number is obtained by doubling the previous number. There are a variety of applications of binary numbers. Many of these applications provide solutions to games and puzzles, such as the popular Mind Reading Cards and the well-known game of "Nim."

1. **Mind Reading Cards:** With these five cards you can determine the age of any person who is not over 31. Which of these cards is your age on? If you selected Card 4 and Card 16, then you are 20 years old. If you selected Cards 1, 2, and 16, then you are 19 years old.

Card 16	Card 8	Card 4	Card 2	Card 1
16 21 26 17 22 27 18 23 28 19 24 29 20 25 30 31	8 13 26 9 14 27 10 15 28 11 24 29 12 25 30 31	4 12 22 5 13 23 6 14 28 7 15 29 20 30 21 31	2 11 23 3 14 26 6 15 27 7 18 30 10 19 31 22	1 11 21 3 13 23 5 15 25 7 17 27 9 19 29 31

a. Each number in the upper left-hand corner of these cards is a binary number. Write your age as the sum of the fewest possible binary numbers. On which cards do the binary numbers for your age appear? On which cards does your age appear?

★ b. Write the number 27 as a sum of binary numbers. On which cards do these binary numbers occur? On which cards does 27 occur?

c. If someone chooses a number that is only on Cards 1, 2, and 8, what is this number?

★ d. Explain how the Mind Reading Cards work.

e. The Mind Reading Cards can be extended to include greater numbers. The next card is Card 32. It is included on Material Card 8. Use this card to extend your system of cards. List all the numbers that should be put on this card. (*Hint:* Look for patterns on the first five cards.)

★ f. To extend this system to six cards, more numbers must be placed on the first five cards. For example, 33 = 32 + 1, so 33 must be put on Card 1 as well as Card 32. On which cards should 44 be placed?

g. Write the additional numbers that must be put on Cards 1, 2, 4, 8, and 16 to extend this system to six cards.

Card 32: 32

Card 16	Card 8	Card 4	Card 2	Card 1
16 21 26 17 22 27 18 23 28 19 24 29 20 25 30 31	8 13 26 9 14 27 10 15 28 11 24 29 12 25 30 31	4 12 22 5 13 23 6 14 28 7 15 29 20 30 21 31	2 11 23 3 14 26 6 15 27 7 18 30 10 19 31 22	1 11 21 3 13 23 5 15 25 7 17 27 9 19 29 31

Transfer the numbers from the six cards in parts **e** and **g** to Material Card 8. Cut out the Mind Reading Cards and use them to intrigue your friends or students. Ask someone to select the cards containing their age, or to pick a number less than 64 and tell you which cards it is on.

ACTIVITY SET 2.3

2. **Game of "Nim"**: The game of "Nim" is said to be of ancient Chinese origin. It is a two-person game and in its simplest form is played with three rows of sticks or markers. On a player's turn any number of sticks may be taken from any one row. The winner of the game is the player who takes the last stick or group of sticks. Try a few rounds of this game with a classmate.

   a. With a little practice it is easy to discover some situations in which you can win. For example, if it is your opponent's turn to play and there is a 1-2-3 arrangement as shown here, you will be able to win. Sketch three more arrangements for which you will win if it is your opponent's turn to play.

   |
   | |
   | | |

   ★ b. *Winning Strategy:* A winning strategy for the game of "Nim" involves the binary numbers. On your turn, group the remaining sticks in each row by binary numbers, using the largest binary numbers possible in each row. Then remove sticks so that there is an *even number* of each type of pile left. In this example you will see an even number of 4s, an even number of 2s, and an odd number of 1s. So one stick must be taken from one row. Show that the 1-2-3 arrangement above has an even number of binary groups.

   c. Assume it is your turn and you have the arrangement shown here. Cross out stick(s) so that your opponent will be left with an even number of binary groups.

   | | | | |
   | | | |
   | |

   ★ d. After the markers in each row have been grouped by binary numbers, there will be either an *even situation* (an even number of each type of pile) or an *odd situation*. Experiment to see whether one turn will always change an even situation to an odd situation. Is it possible in one turn to change an even situation into another even situation?

   e. *Generalized "Nim":* Play the game of "Nim" with any number of rows and any number of sticks in each row. Try the strategy of leaving your opponent with an even situation each time it is his or her turn. Is this still a winning strategy?

＃ 3

Whole Numbers and Their Operations

ACTIVITY SET 3.1

ADDING WITH MULTIBASE PIECES AND CHIP TRADING

"But don't panic. Base 8 is just like base 10 really, if you're missing two fingers."

Tom Lehrer

One of the best ways of obtaining an insight into our algorithm for addition is by thinking through this process in other bases. The "Trading Up Game," which is explained with multibase pieces in Activity 1 and Chip Trading in Activity 8, provides a natural introduction to the addition of numbers in various bases. Material Cards 31, 11, and 36 contain multibase pieces, a chip mat, and chips. Cut out the multibase pieces and chips for these activities.

1. **"Trading Up Game"* (2-5 players):** On a player's turn two dice are rolled. The product (or the game can be played with sums) of the two numbers from the dice is the number of units the player wins. At the end of a player's turn the pieces won should be traded (regrouped) so that the total winnings are represented by the fewest possible number of base five multibase pieces. The first player to get 1 block or more wins the game.

   Example On a player's first turn, two 6s were rolled on the dice. The product of 36 is represented by 1 flat, 2 longs, and 1 unit.

   Four longs and 4 units were won on the player's second turn. What was the product from the dice?

   This player now has a total of 1 flat, 6 longs, and 5 units. The turn is finished by trading in (regrouping) 5 units for 1 long, and 5 longs for 1 flat. Two flats and 2 longs are the minimum number of pieces that can be used to represent the player's total score at the end of his/her second turn.

   Use your base five multibase pieces to play this game.

★ 2. The following tables show the numbers of base five pieces that were won in the "Trading Up Game" by three players. After each player had seven turns the game ended. Determine the base five pieces that each player had at the end of the game. Who won the game? (You may find it helpful to use your base five multibase pieces to carry out the trading for the turns of each player.)

*This game is adapted from P.S. Davidson, G.K. Galton, and A.W. Fair, *Chip Trading Activities* (Fort Collins, Colorado: Scott Resources, 1970).

Player 1				Player 2				Player 3			
B	F	L	U	B	F	L	U	B	F	L	U
		3	3			2	0			1	0
		4	4			2	2			2	2
		3	0			4	0		1	0	0
		2	2			3	3			3	1
		4	4			4	0		1	1	0
		4	0		1	1	0			3	3
		2	2			2	0			1	1
Total				Total				Total			

Player 1 got 3 longs and 3 units on the first turn. Therefore, 3 and 6 must have been rolled on the dice. Explain why. Determine the product from the dice that will give the remaining base five pieces that Player 1 received.

3. The addition of numbers written with base five numerals is similar to determining the minimum numbers of units, longs, and flats in the "Trading Up Game." Here are the base five pieces for the numerals 324_{five} and 243_{five}. To compute this sum, 5 of the 7 units were regrouped into 1 long. What other regroupings are necessary?

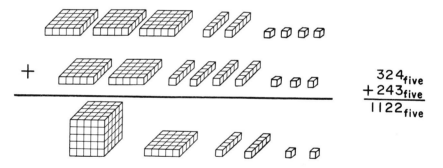

$$\begin{array}{r} 324_{five} \\ + 243_{five} \\ \hline 1122_{five} \end{array}$$

Use your base five pieces to compute these sums.

★ a.   BFLU
    1304 $_{five}$
+   203 $_{five}$

b.   BFLU
    1402 $_{five}$
+ 2344 $_{five}$

★ c.   BFLU
    1103 $_{five}$
     232 $_{five}$
+ 2144 $_{five}$

4. Here are the 4 different types of multibase pieces for base eight.

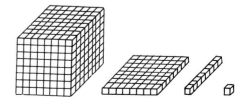

After five rounds of using the base eight multibase pieces in a game of "Trading Up," the players had recorded the scores in the following table. Find the total score for each player in terms of the minimum number of base eight pieces. Who is the closest to winning a block?

Player 1					Player 2					Player 3					Player 4			
B	F	L	U	B	F	L	U	B	F	L	U	B	F	L	U
		3	1			2	4			1	7			4	4
		1	2				6			2	4				3
		1	7			1	4			2	2			1	2
		2	4			3	6			3	0			3	1
		3	0			1	4			2	0			1	2
Total				Total				Total				Total			

★ 5. Use these pictures of base eight pieces to compute $1436_{eight} + 257_{eight}$. In this example you will have to regroup 8 units into 1 long, leaving 5 units. What other regroupings are necessary? After regrouping, sketch the multibase pieces for this sum.

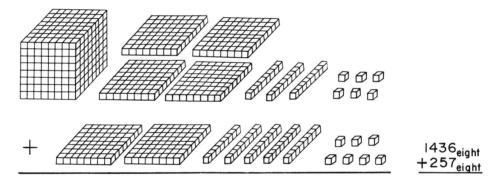

$$\begin{array}{r} 1436_{eight} \\ +257_{eight} \\ \hline \end{array}$$

6. What base are these multibase pieces used for?

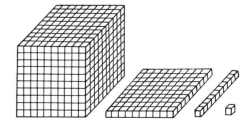

34 CHAP. 3 WHOLE NUMBERS AND THEIR OPERATIONS

In a "Trading Up Game" with these multibase pieces, a player got the following pairs of numbers from the dice on 4 turns: 6,6; 4,5; 6,5; and 4,6. Write the numbers of longs and units for the products of these numbers in the accompanying table. Then determine the minimum numbers of flats, longs, and units after the player's fourth turn.

	B	F	L	U
First turn				
Second turn				
Third turn				
Fourth turn				
Total				

7. To add in various bases we can visualize the digits as numbers of multibase pieces. The blocks, flats, longs, and units are shown for bases three, four, six, and ten in the following exercises. Compute these sums.

★ a.   BFLU
       1221_{three}
   + 122_{three}

b.   BFLU
       2312_{four}
   + 203_{four}

★ c.   BFLU
       340_{six}
       255_{six}
  + 1133_{six}

d.   BFLU
       8645
  + 1776

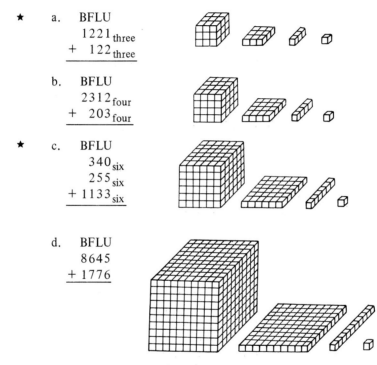

8. **"Trading Up Game"—Base Five (Chip Trading Mat and Chips, Material Cards 11 and 36):** Each player begins with a mat having no chips on it. On a player's turn two dice are rolled and the product (or the game can be played with sums) of the two numbers is the number of yellow chips the player wins. At the end of a player's turn the chips should be traded so that each group of 5 chips on a column is replaced by 1 chip on the column to the left. The first player to get 1 red chip wins the game.

Red	Green	Blue	Yellow
		⊙⊙⊙	⊙⊙⊙

The chips on the above mat were obtained from an 18 on the first turn.

Play this game with someone and record the numbers of green, blue, and yellow chips that are won on each of your turns.

a. How many turns did the winning player require?

★ b. A red chip is worth how many yellow chips?

★ c. What is the smallest possible number of turns required for obtaining 1 red chip in this game?

d. Use the accompanying table to determine the numbers of green, blue, and yellow chips you had after your fifth turn.

Record of Turns

Turns	Green	Blue	Yellow
1			
2			
3			
4			
5			
6			
7			
8			
9			
10			
11			

9. **"Trading Up Game"—Base Eight**: In this game, 8 yellow chips trade for 1 blue chip, 8 blue chips for 1 green chip, and 8 green chips for 1 red chip.

★ a. A red chip is worth how many yellow chips?

★ b. What is the minimum number of turns required to win 1 red chip?

c. Player A and Player B obtained the products shown above the following mats. Use your mat and chips to determine the numbers of each type of chip that they had after their eleventh turn. Indicate your answer by drawing chips on the mats. Which player is closest to winning a red chip?

Player A

12, 15, 1, 6, 5, 5,
36, 24, 24, 18, 12

Red	Green	Blue	Yellow

Player B

6, 2, 30, 20, 4, 5,
4, 3, 9, 25, 24

Red	Green	Blue	Yellow

10. Placing numerals on a Chip Trading mat provides an intermediate step between using chips and the standard algorithm for addition. Consider adding 1565_{eight} and 2734_{eight} on the mat shown here. If we were using chips, in the first step of the algorithm 8 yellow chips would be traded for 1 blue chip, and 1 yellow chip would remain. To indicate the regrouping, 1 chip has been placed in the second column. Finish computing this sum.

Base Eight

Red	Green	Blue	Yellow
		⊙	
1	5	6	5
+2	7	3	4
☐	☐	☐	1

Compute the following sums for the given bases. (You may find it helpful to use the chips for regrouping.)

★ a. Base Seven b. Base Three ★ c. Base Six

Red	Green	Blue	Yellow
3	6	2	5
+1	2	4	6
☐	☐	☐	☐

Red	Green	Blue	Yellow
1	2	0	2
+	1	2	1
☐	☐	☐	☐

Red	Green	Blue	Yellow
2	0	5	5
+1	4	2	3
☐	☐	☐	☐

ACTIVITY SET 3.2

SUBTRACTING WITH MULTIBASE PIECES

The operations of subtraction and addition are inverses of each other. *Addition* is explained by "putting together" sets of objects, and *subtraction* is explained by "taking away" a subset of objects from a given set. This dual relationship between subtraction and addition can be seen in the "Trading Up Game" (Activity Set 3.1) and the "Trading Down Game" in the following activities. This game is played with the base five multibase pieces from Material Card 31. The trading of multibase pieces is similar to the regrouping or "borrowing" in the subtraction algorithm. The multibase pieces will be used to illustrate this algorithm and to develop a "new" algorithm for subtraction.

★ 1. **"Trading Down Game"**: Each player begins the game with 1 block. On a player's turn, two dice are rolled. The product of the numbers from the dice determines the number of units the player discards. The object of the game is to get rid of all your base five pieces. The first player to do so wins the game. On some turns, and in particular the first turn of the game, each player will have to do some trading (regrouping). At the end of each turn the remaining units should be represented by the fewest possible number of base five pieces.

Example On a player's first turn a product of 18 means that he/she must trade away 3 longs and 3 units. Use your base five multibase pieces to determine the numbers of units, longs, and flats after the first turn. Write these numbers in the table.

Use your base five multibase pieces to play this game.

	B	F	L	U
Beginning of game	1	0	0	0
First turn			3	3
After first turn				

★ 2. The top rows of the following tables show which base five pieces the players had left after 3 turns in the "Trading Down Game." The second rows show the numbers of base five pieces each player has to get rid of on the fourth turn. Determine the numbers of flats, longs, and units each player will have after 4 turns.

Player 1

B	F	L	U
	3	2	3
		4	0

Player 2

B	F	L	U
	4	1	1
		4	4

Player 3

B	F	L	U
	3	1	4
	1	2	1

Player 4

B	F	L	U
	2	2	2
		3	3

3. The base five multibase pieces are used here to illustrate $422_{five} - 243_{five}$. In order to remove the 3 circled units, the multibase pieces for 422_{five} have been regrouped by replacing 1 long by 5 units. Then to subtract the 4 circled longs, 1 flat has been replaced by 5 longs. When 2 flats, 4 longs, and 3 units have been removed, the remaining pieces represent the difference, 124_{five}.

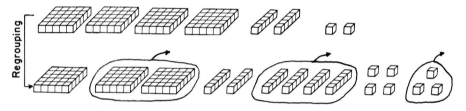

Use your multibase pieces to compute these differences.

★ a. BFLU
        1142_{five}
      $-\ 213_{five}$

b. BFLU
      2331_{five}
    $-\ 124_{five}$

★ c. BFLU
        4112_{five}
      $-\ 143_{five}$

★ 4. Here are the 4 types of base seven multibase pieces. On the second turn of "Trading Down" using these pieces, the players had recorded the scores in the following tables. Determine the numbers of flats, longs, and units each player will have after the second turn.

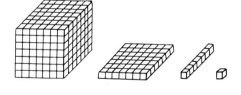

	Player 1				Player 2				Player 3				Player 4			
	B	F	L	U	B	F	L	U	B	F	L	U	B	F	L	U
After first turn		6	6	1		6	5	2		6	4	1		6	6	4
Second turn			2	6			3	3			5	1				6
After second turn																

5. Use these base seven pieces to compute $2034_{seven} - 1235_{seven}$. Cross out and sketch multibase pieces to show the necessary regrouping. Encircle the pieces that should be taken away to illustrate the difference.

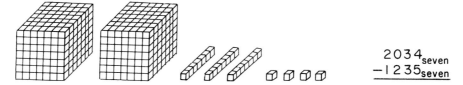

$$\begin{array}{r} 2034_{seven} \\ -1235_{seven} \\ \hline \end{array}$$

★ 6. The following tables show the block that each player had at the beginning of the "Trading Down Game" with base ten pieces and the scores on their first turns. What are the numbers of flats, longs, and units each player will have after the first turn?

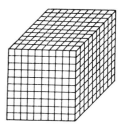

	Player 1				Player 2				Player 3			
	B	F	L	U	B	F	L	U	B	F	L	U
Beginning of game	1	0	0	0	1	0	0	0	1	0	0	0
First turn			1	8			2	4			3	6
After first turn												

7. To subtract in various bases we can visualize the digits as numbers of multibase pieces. The blocks, flats, longs, and units are shown for bases three, four, and eight in the following exercises. Compute these differences.

★ a. $\begin{array}{r} \text{BFLU} \\ 2121_{three} \\ -\ 122_{three} \\ \hline \end{array}$

b. $\begin{array}{r} \text{BFLU} \\ 3202_{four} \\ -1133_{four} \\ \hline \end{array}$

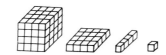

★ c. $\begin{array}{r} \text{BFLU} \\ 4032_{eight} \\ -2316_{eight} \\ \hline \end{array}$

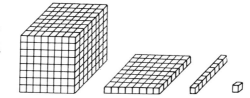

ACTIVITY SET 3.2

8. The main purpose of using multibase pieces (bundles of sticks, Chip Trading, abacus, etc.) for computing is to provide an understanding of the operations. Sometimes, however, rather than reinforce the standard approaches to the basic operations, these manipulatives can lead to new algorithms. Consider the following example which uses the base ten multibase pieces to compute 2493 − 826.

2493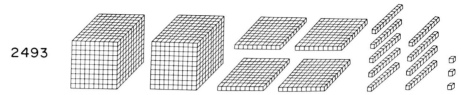

Step 1 (Subtracting 6 units): Since 3 is less than 6, we can take 6 units from 1 long. This leaves 4 units, which are placed with the 3 units to make 7 units. The number of longs has been decreased from 9 to 8.

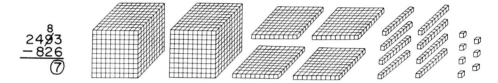

Step 2 (Subtracting 2 tens): Since 8 is greater than 2, we can take away 2 longs from the 8 longs, leaving 6 longs.

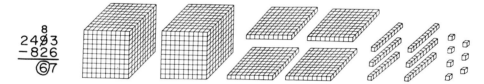

Step 3 (Subtracting 8 hundreds): Since 4 is less than 8, we take 8 flats from 1 block. This leaves 2 flats, which are placed with the 4 flats to make 6 flats. The number of blocks has been decreased from 2 to 1.

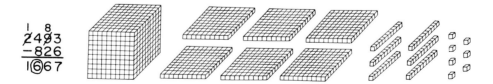

Use this method on the following multibase pieces to compute the given differences. Show the results of subtracting by encircling pieces and show regrouping by sketching in new pieces.

a. 542 − 217

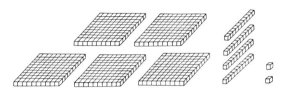

b. 2129 − 654

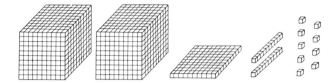

9. The previous approach to subtraction leads to a new algorithm.* In this algorithm digits are subtracted in the same column, as usual, when no borrowing is necessary. When borrowing is needed the subtracting is done by using the digit representing the next higher power of ten (as in the preceding examples with multibase pieces). Use this method to compute the following differences. Name an advantage of this algorithm over the usual (decomposition) method of subtracting.

★ a. 3471   b. 7235   c. 4092
   − 128   − 473   − 347

ACTIVITY SET 3.3

MULTIPLYING WITH THE ABACUS AND CHIP TRADING

The abacus has been used for many centuries to compute with the four basic operations. Howard W. Eves describes a computing contest between a Japanese clerk with an abacus and an American Army clerk with an electric desk calculator.** The contest, which was held in Tokyo, was witnessed by some 3000 spectators. The abacus was a Japanese soroban with a prewar value of 25 cents, and the desk calculator was valued at $700. The Japanese clerk fluttered his hand over his soroban with such blurring dexterity that the Americans nicknamed him "The Hands." "The Hands" won all six heats of the addition event, finishing one of them a minute ahead of the army clerk. "The Hands" lost in multiplication, but he won in subtraction, division, and in the final composition problem involving all four operations.

---

*This algorithm is described by H. Ikeda and M. Ando, "A New Algorithm for Subtraction," *The Arithmetic Teacher,* **21** No. 8 (December 1974), pp. 716–19.
**H.W. Eves, *Mathematical Circles Revisited* (Boston: Prindle, Weber and Schmidt, 1971) p. 37.

Multiplication is carried out in the following activities with the abacus and Chip Trading. Bases other than ten are used in some activities to encourage thinking through the concepts of multiplying and regrouping. There are variable base and base ten abacuses on Material Cards 7 and 13, and markers for these abacuses are on Material Card 30. You may find it helpful to think of the powers of the bases in terms of units, longs, flats, and blocks. These multibase pieces are pictured on your variable base abacus. A mat and chips are on Material Cards 11 and 36 for the Chip Trading activities.

1. Our algorithm for multiplication is a process by which the products of large numbers can be obtained by using only the products of single-digit numbers. This is illustrated in the computation of $3 \times 1432_{\text{five}}$ on the variable base abacus. On this abacus the number of markers on each column for 1432_{five} has been tripled. Regroup these markers and sketch the results on the second abacus.

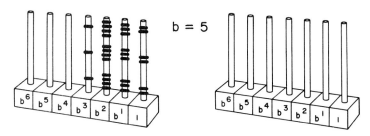

Use your variable base abacus to compute these products.

★ a. $4 \times 342_{\text{five}}$ b. $3 \times 203_{\text{four}}$ ★ c. $2 \times 6377_{\text{eight}}$

2. In base five, 5 is represented by 10_{five}. To multiply by 10_{five}, each marker on the first abacus has been replaced by 5 markers on the second abacus. Regroup these markers and show the results on the third abacus.

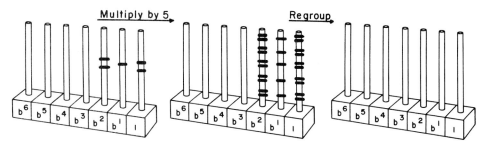

★ a. Use your variable base abacus to compute the following products.
   $10_{\text{five}} \times 1234_{\text{five}} =$ $10_{\text{five}} \times 203_{\text{five}} =$

★ b. State a rule that gives the results of multiplying in base five by 10_{five}.

  c. Will this rule hold for multiplying in base eight by 10_{eight}?

★ 3. The product 3 × 435 can be computed on your base ten abacus (Material Card 13) by tripling the number of markers in each column for the number 435. Since 10 times 100 is equal to 1000, 10 markers from the hundreds column can be replaced by 1 marker on the thousands column. What other regroupings are possible? Regroup these markers and show the results of your regrouping on the second abacus.

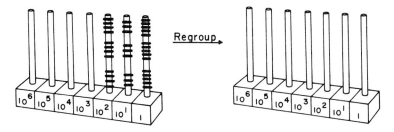

The algorithm for multiplying 3 × 435 is shown next in two stages. In the first stage the digits of 435 were multiplied by 3. In the second stage the regrouping was done.

```
   Stage 1            Stage 2
   | 4|3|5           |  |4|3|5
   |  |X|3           |  | |X|3
   |12|9|15          |1 |3|0|5
```

Compute 4 × 3041 using your base ten abacus. First multiply the number of markers in each column by 4 and then regroup. Record your results in the following two stages of the algorithm.

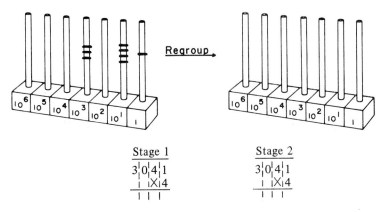

```
   Stage 1            Stage 2
   3|0|4|1           3|0|4|1
    | |X|4            | |X|4
    | | |             | | |
```

4. To compute 7 × 6523, the multiplying and regrouping have been done in one stage, as shown on the second abacus. The regrouping was done as the number of markers in each column of the first abacus was multiplied by 7. For example, beginning with the units column, 7 × 3 = 21, so 1 marker was put on the units column of the second abacus. The 20 was recorded by placing 2 markers at the top of the tens column.

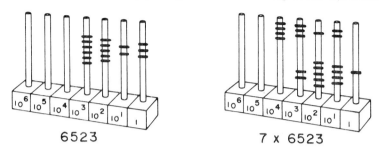

★ a. There are markers at the tops of the 10^2, 10^3, and 10^4 columns, which came from regrouping. Explain what products each of these groups of markers came from.

b. Do the 2 markers at the top of the tens column on the second abacus get multiplied by 7? Give a reason for your answer.

★ c. Use the algorithm for multiplication to compute 7 × 6523. Compare each step of the algorithm with those used on the abacus. Which numbers that occur in the algorithm correspond to the markers at the tops of the columns on the abacus?

```
 6523
 × 7
```

5. Use your base ten abacus to compute the following products. Then use the algorithm for multiplication and compare the steps in each process.

★ a. 30 × 271. This product can be computed by multiplying first by 3 and then by 10, since 30 × 271 = (10 × 3) × 271 = 10 × (3 × 271). What number property is used in this equation?

```
 271
 ×30
```

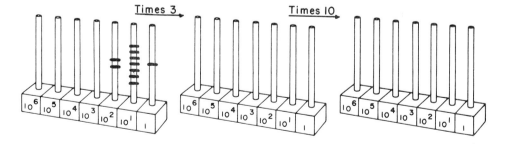

44   CHAP. 3   WHOLE NUMBERS AND THEIR OPERATIONS

b. 200 × 4831. This product can be computed by first multiplying by 2 and then by 100, since 200 × 4831 = (100 × 2) × 4831 = 100 × (2 × 4831).

```
  4831
×  200
```

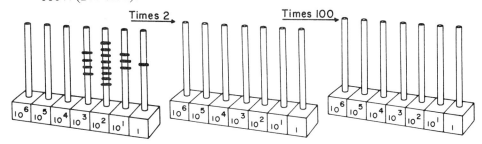

6. To compute 34 × 132, we can compute 4 × 132 plus 30 times 132, since (30 + 4) × 132 = 30 × 132 + 4 × 132. What number property is used in this equation? Compute each of these products and show the results by drawing markers on the abacuses. Add the numbers represented by the markers you have drawn to get the product.

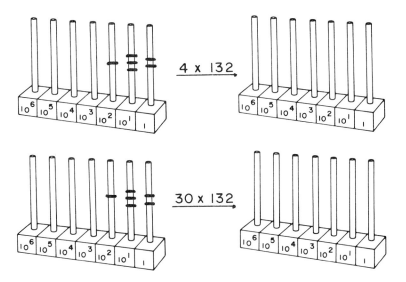

★ a. Compare the steps used on the abacus with those used in the algorithm shown here. In which column of the abacus will the markers for the encircled 6 be?

★ b. The "396" in the algorithm represents 3960. What is there about the use of the abacus which might account for the practice of not writing the "0" in "3960" for the algorithm?

```
   132
 × 34
  528
  396
 4488
```

ACTIVITY SET 3.3   45

7. **Multiplying with Chip Trading**: The Chip Trading model for multiplication is essentially the same as the methods we have used for multiplying on the abacus. First, there is the process of multiplying and then there is the regrouping. The chips on the mat on the left were obtained from multiplying 342_{five} by 4. Finish computing this product by doing the regrouping. Sketch your results on the second mat.

Base Five

Red	Green	Blue	Yellow
		⊙	
⊙⊙⊙	⊙⊙⊙	⊙⊙⊙	⊙⊙
⊙⊙⊙	⊙⊙⊙	⊙⊙⊙	⊙⊙
⊙⊙⊙	⊙⊙⊙	⊙⊙⊙	⊙⊙
⊙⊙⊙	⊙⊙⊙	⊙⊙⊙	⊙⊙

Regroup →

Red	Green	Blue	Yellow

Use your mat and chips from Material Cards 11 and 36 to compute these products.

★ a. $2 \times 637_{\text{eight}}$ b. $2 \times 212_{\text{three}}$ ★ c. $3 \times 404_{\text{five}}$

8. Products by multiples of 10 can be accomplished in two stages. Sketch chips on the following mats to show the results of multiplying and regrouping for each of the given steps.

★ a. 30×284

Red	Green	Blue	Yellow
		⊙⊙	⊙⊙
	⊙⊙	⊙⊙	⊙⊙
		⊙⊙	
		⊙⊙	

Times 3 →

Red	Green	Blue	Yellow

Times 10 →

Red	Green	Blue	Yellow

b. 60×107

Red	Green	Blue	Yellow
			⊙
	⊙		⊙⊙
			⊙⊙
			⊙⊙

Times 6 →

Red	Green	Blue	Yellow

Times 10 →

Red	Green	Blue	Yellow

9. Numerals rather than chips can be placed on the mat to provide an intermediate step between computing with chips and using the algorithm for multiplication. For example, consider the product 6 × 735$_{eight}$. If we were using chips, there would be 6 × 5 or 30 yellow chips. Regrouping in base eight, this is equivalent to 3 blue chips and 6 yellow chips left over. The numeral 6 is used for 6 yellow chips, and the regrouping is indicated by placing 3 blue chips in the second column (see mat). Finish computing this product. Draw green and red chips to indicate the remaining regrouping.

Base Eight

Red	Green	Blue	Yellow
		⊙⊙⊙	
7	3	×	5 6
			6

Compute the following products for the given bases. Sketch in chips to indicate the regrouping process.

★ a. Base Four

Red	Green	Blue	Yellow
3	3	3	
		×	3

b. Base Nine

Red	Green	Blue	Yellow
	8	4	6
		×	7

★ c. Base Three

Red	Green	Blue	Yellow
	2	1	2
		×	2

Just for Fun—Cross-numbers for Calculators

Sit back, relax, and exercise your calculator's multiplication key on this cross-number puzzle. Not all answers are obtainable on a calculator, however. Clues for some of the numbers have been omitted, but no square is left blank.

ACROSS

1. [(18 + 2) × 4] + 19
5. $\dfrac{14 + 3}{9 + 8} \times 66$
6. 500,000,000 ÷ 9 (omit the decimal)
8. The middle two digits of the product of 2 × 2 × 2 × 2 × 3 × 37
12. $\dfrac{(41 - 29) \times (63 - 23)}{95 \div 19}$
15. The sum of the digits in the product of 13 × 9004 × 77
16. The sum of the first seven odd numbers

(Continued on next page)

DOWN

3. (34 × 53) − (137 × 2)
4. Asphalt roof shingles are sold in bundles of 27. Three bundles cover 100 sq. ft. How many bundles are needed for 11,700 sq. ft. of roofing in a housing development?

(Continued on next page)

ACROSS (Cont.)

17. March 30, 1974: A "freeze model" in a department store posed motionless for 5 hr., 32 min. How many minutes in all did he pose?
23. $\dfrac{13334 - 6566}{8 \times 9} \times \dfrac{992 \div 4}{17 + 14}$
27. $(10{,}000{,}000 \div 81) \times 100$ (nearest whole number)
28. Take current year, subtract 17, multiply by 2, add 6, divide by 4, add 21, subtract 1/2 of current year.
29. First two digits of the product of 150×12

DOWN (Cont.)

5. $\dfrac{88984 \times 493}{17 \times 392}$
7. $\dfrac{51 \times 153 \times 62 \times 57}{17 \times 969 \times 31}$
11. $\dfrac{(62 \times 21) + 23}{25} - 11$
14. The number of different ways to change a dollar using all American coins is a palindrome. The sum of its digits is 13.
17. Remainder of $1816 \div 35$
18. $(343 \times 5 + 242) \times 2$
19. $4 \times [(37 \times 24) - 1]$
21. The number halfway between 7741 and 8611
26. $\dfrac{119 \times 207 \times 415}{23 \times 747}$

ACTIVITY SET 3.4

DIVIDING ON THE ABACUS

Dividing by repeated subtraction is called the *measurement concept of division*. This approach to division is the one most often used to explain the long division algorithm and is the method illustrated on the abacus in the following activities. Most of these activities involve the base ten abacus (Material Card 13) and the similarities between using this abacus and the long division algorithm. The variable base abacus (Material Card 7) is also used to show that the concept of division is not affected by changes in the base.

1. The markers on this abacus represent 693. To compute $693 \div 3$, remove (measure off) 3 markers at a time from a column. Record the number of groups of 3 by placing a marker above the column for each group. For example, for the removal of the 3 encircled markers on the hundreds column, 1 marker is placed above this column.

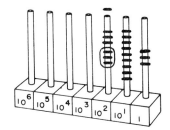

★ a. If 3 markers are removed from the hundreds column, how many 3s have been subtracted from the original number?

★ b. What does the one marker shown above the hundreds column represent?

★ c. To compute $693 \div 3$ on the abacus, does it matter which column you begin with?

2. Form the number 5264 on your base ten abacus and compute 5264 ÷ 4. When there are less than 4 markers in a column, regroup by replacing a marker from a column by 10 markers on the column to the right.

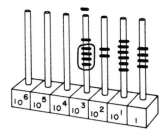

★ a. If the 4 encircled markers are removed from the 10^3 column, 1 marker should be placed above this column. Why?

★ b. Explain what is done with the remaining marker in the 10^3 column.

c. Check your answer from the abacus by using the long division algorithm to compute 5264 ÷ 4. When 4 markers are removed from the 10^3 column, 1 marker remains. Encircle the number in the algorithm that corresponds to this marker.

d. On the abacus we can divide with the markers in one column at a time. Is it just as convenient to begin with the units column as with the column on the left? Try it for the preceding example.

★ 3. To compute 1212 ÷ 3 on the base ten abacus, markers can be regrouped from left to right so that 3 markers can be removed from a column at a time. Form this number on your base ten abacus and compute the quotient. Begin by replacing the marker on the 10^3 column by 10 markers on the 10^2 column. When you have finished dividing on the abacus, use long division to compute 1212 ÷ 3. Use arrows to connect each group of markers above the columns with the corresponding digits in the quotient of the division algorithm.

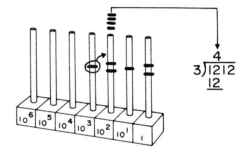

Compute the quotient 8543 ÷ 7 using both your base ten abacus and the long division algorithm. Match each group of markers above the columns with the corresponding digits in the quotient of the algorithm.

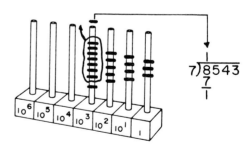

ACTIVITY SET 3.4   49

4. Changing the base does not change the concept of division. By letting $b = 7$, the variable base abacus can be used to compute $603_{seven} \div 3$. Each time 3 markers are removed from a column, 1 marker is put above that column. For example, 2 markers have been placed above the 7^2 column because $(6 \times 7^2) \div 3 = 2 \times 7^2$. What is $603_{seven} \div 3$ equal to in base seven?

Use your variable base abacus to compute these quotients.

★ a. $603_{eight} \div 3 =$ b. $4204_{five} \div 2 =$ ★ c. $8044_{nine} \div 4 =$

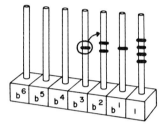

★ 5. To regroup in base seven, each marker from a column is replaced by putting 7 markers on the column to its right. Use your variable base abacus to compute $1214_{seven} \div 4$. Do the necessary regrouping so that 4 markers at a time can be removed from a column.

Use your variable base abacus to compute the following quotients. Whenever the number of markers on a column is less than the divisor, regroup.

★ a. $1224_{six} \div 4 =$ b. $2121_{three} \div 2 =$ ★ c. $1665_{seven} \div 6 =$

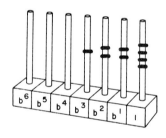 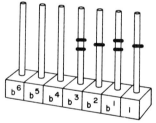 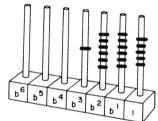

6. In the previous examples you have been instructed to regroup the markers whenever the number of markers on a column is less than the number you are dividing by. There is a more convenient method of dividing on the abacus, which is closer to the algorithm for long division. This method will be used to compute $276 \div 4$. The 2 markers on the hundreds column and 4 of the markers on the tens column of

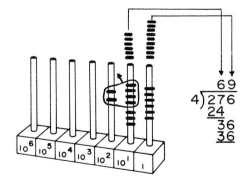

this abacus represent 24 tens. Therefore, we can subtract off these markers and place 6 markers over the tens column to indicate part of the quotient.

$(10 \times 24) \div 4 = 10 \times (24 \div 4) = 10 \times 6$

The remaining markers on the tens and units columns represent 36. Since $36 \div 4 = 9$, we place 9 markers above the units column. The quotient, 69, is represented by the markers above the tens and units columns.

Compute the following quotients by using markers on two columns at a time. Encircle the appropriate groups of markers and sketch markers above the columns to represent the quotient. Compute these quotients by the long division algorithm and match the markers above the columns with the corresponding digits in the division algorithm.

★ a. $441 \div 7$ b. $2538 \div 6$

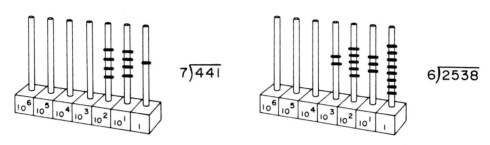

7. Represent 7944 on your base ten abacus and compute $7944 \div 6$. Whenever there are fewer than 6 markers on a column, use the markers on two adjacent columns, as in Activity 6. Compare the steps used on the abacus with those in the long division algorithm.

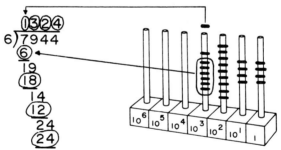

Every number appearing in the long division algorithm will correspond to a set of markers when you compute this quotient on your abacus. For example, the encircled 6 has the value 6000. This was subtracted from 7944 by removing 6 markers from the 10^3 column. The circled 1 has the value 1000 and is represented by the marker at the top of the 10^3 column. Give the value of the other circled numbers in the algorithm and explain how they are related to the computation on the abacus.

⑥ has the value 6000. Thus, 6 markers were removed from the thousands column.

① has the value 1000. Thus, 1 marker was placed above the thousands column.

⑱

ACTIVITY SET 3.4 51

③
⑫
②
㉔
④

Just for Fun—Calculator Games and Number Tricks

Arabian Nights Mystery: Select any three-digit number, such as 837, and enter the six-digit number 837837 on your calculator. Then carry out the following steps: Divide this number by 11; then divide the result by 7; and finally, divide the result by 13. You may be surprised by the result. This trick has been called the "Arabian Nights Mystery" because it can be explained by using the number 1001. (This is in reference to the book, *Tales from the Thousand and One Nights*.) Try this trick for some other three-digit numbers. Explain why it works.

"Keyboard Game" (2 players): The game begins by one player entering a whole number on a calculator. The players then take turns subtracting single-digit numbers, with the following restriction: Whatever digit is subtracted on a given turn, the next player must subtract a nonzero digit from an adjacent button. For example, if a player subtracts 6, the next player must subtract 9, 8, 5, 2, or 3. When a player's turn results in a number less than 0, that player loses the game.

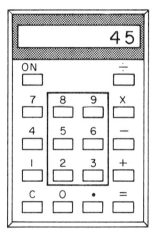

Magic Formulas: Usually if someone asks you to select a number to perform some operations on, you will choose an easy one to work with. However, with the aid of a calculator you do not have to be so careful about the number you select. For example, try a three- or four-digit number in the following formula.

Select any number (remember this number, as it will be needed later); add 221 to your number; multiply by 2652; subtract 1326; divide by 663; subtract 870; divide by 4; subtract the number you started with. Your result will be 3.

Most magic formulas such as the one preceding require that the original number be subtracted off in the computation. There is no such requirement in the following formula. This trick, however, will work only on calculators with an eight-digit display and no "hidden digits." If your calculator has more places for digits, design a similar formula which can be used on your calculator.

Select a three-digit number and divide it by 9; divide this result by 10,000,000; multiply by 1,000,000; add 300; divide by 10,000,000; divide by 100; multiply by 10,000,000. Your result will be 3.

ACTIVITY SET 3.5

PATTERNS ON GRIDS

Several rectangular grids of different sizes will be used to locate multiples and factors of numbers. The multiples of a number, such as the multiples of 3 (3, 6, 9, 12, 15, ...), form patterns that vary with the size of the grid. The sieve of Eratosthenes, a method of locating primes by crossing out multiples of numbers, will be used to find primes that are less than 200. The last two grids in this activity set are formed by spirals of numbers and contain diagonal patterns of primes.*

★ 1. One of the most common grids for finding multiples of whole numbers contains ten numbers in a row. There are many number patterns in this type of grid. If this grid were continued, in what row and column would 437 appear?

(a)

1	2	3	4	5	6	7	8	9	10
11	12	13	14	15	16	17	18	19	20
21	22	23	24	25	26	27	28	29	30
31	32	33	34	35	36	37	38	39	40
41	42	43	44	45	46	47	48	49	50
51	52	53	54	55	56	57	58	59	60
61	62	63	64	65	66	67	68	69	70
71	72	73	74	75	76	77	78	79	80
81	82	83	84	85	86	87	88	89	90
91	92	93	94	95	96	97	98	99	100
101	102	103	104	105	106	107	108	109	110
111	112	113	114	115	116	117	118	119	120

a. Draw a circle around each multiple of 3. Let's refer to this set of numbers by $m(3)$. These multiples form a pattern of "left one, down one." How else can this pattern be described?

★ b. Draw a square around each multiple of 4 in grid (a). Describe the pattern formed by $m(4)$.

*M. Gardner, *Sixth Book of Mathematical Games from Scientific American* (New York: Charles Scribner's Sons, 1971), "Patterns and Primes," pp. 79–90.

c. Draw a circle around each multiple of 7 in grid (b). Describe the pattern formed by $m(7)$.

★ d. Describe the patterns formed by $m(9)$ and $m(11)$.

(b)

1	2	3	4	5	6	7	8	9	10
11	12	13	14	15	16	17	18	19	20
21	22	23	24	25	26	27	28	29	30
31	32	33	34	35	36	37	38	39	40
41	42	43	44	45	46	47	48	49	50
51	52	53	54	55	56	57	58	59	60
61	62	63	64	65	66	67	68	69	70
71	72	73	74	75	76	77	78	79	80
81	82	83	84	85	86	87	88	89	90
91	92	93	94	95	96	97	98	99	100
101	102	103	104	105	106	107	108	109	110
111	112	113	114	115	116	117	118	119	120

2. Use the sieve of Eratosthenes to find the prime numbers that are less than 200. Begin by circling 2 and crossing out the remaining multiples of 2. Continue sieving by circling 3, 5, 7, 11, and 13, and crossing out the remaining multiples of these numbers.

★ a. The numbers greater than 1 which are not crossed out are the primes less than 200. Why is it unnecessary to cross out multiples of the primes greater than 13?

b. What is the greatest number of consecutive composite (nonprime) whole numbers less than 200?

★ c. It is reasonable to expect that there will be fewer and fewer primes in equally spaced intervals as we progress through the number system. However, this has never been proven. Find the number of primes in the following intervals.

Interval	1 to 40	40 to 80	80 to 120	120 to 160	160 to 200
Number of Primes	12				

d. Pairs of primes such as 5,7 and 11,13 which differ by 2 are called *twin primes*. It is not known whether or not there are an infinite number of such primes. However, it has been proven that there are fewer and fewer twin primes as the numbers increase. Find the number of twin primes in each of the following intervals.

Interval	1 to 40	40 to 80	80 to 120	120 to 160	160 to 200
Number of Twin Primes	5				

ACTIVITY SET 3.5

★ 3. Here's an interesting variation of the sieve which uses colors to indicate the multiples of 2, 3, 5, and 7. In each box with a multiple of 7, there is a black corner. Color a corner blue for each multiple of 2, red for each multiple of 3, and green for each multiple of 5. This idea was originated by an elementary school student named Keith.* Originally, Keith made a color-coded sieve for the numbers up to 1000. How many different colors did he use for his grid?

1	2	3	4	5	6	7	8	9	10
11	12	13	14	15	16	17	18	19	20
21	22	23	24	25	26	27	28	29	30
31	32	33	34	35	36	37	38	39	40
41	42	43	44	45	46	47	48	49	50
51	52	53	54	55	56	57	58	59	60
61	62	63	64	65	66	67	68	69	70
71	72	73	74	75	76	77	78	79	80
81	82	83	84	85	86	87	88	89	90
91	92	93	94	95	96	97	98	99	100
101	102	103	104	105	106	107	108	109	110
111	112	113	114	115	116	117	118	119	120

  a. Use these colors to find the different prime factors of 84 (the primes that divide into 84 evenly).

★ b. Use the colors to indicate which numbers in this chart have 2 and 7 as factors.

  c. What prime factors do 84 and 30 have in common?

★ d. Use this chart to find all the numbers which are multiples of both 3 and 5.

  e. Are the numbers in the uncolored cells all prime?

*C. L. Bradford, "Keith's Secret Discovery of the Sieve of Eratosthenes," *The Arithmetic Teacher*, **21** No. 3 (March 1974), pp. 239–41.

4. A grid with six numbers in a row has many features which are not found in the previous grids.* The prime numbers which are normally scattered and unpredictable become easier to locate. Furthermore, you will see that the multiples of the primes form patterns having interesting relationships.

★ a. What pattern do the multiples of 5 form? How many vertical spaces are between adjacent lines of this pattern?

   b. What pattern do the multiples of 7 form? How many vertical spaces are between adjacent lines of this pattern?

★ c. The multiples of 11 and 13 also lie on straight lines. Draw lines through these multiples. What is happening to the patterns of multiples as the primes get greater?

   d. Draw lines through the multiples of 2 and 3. The numbers greater than 1 which remain will be prime. Why was it necessary to cross out only multiples of primes up to 13 to find the primes less than 250?

★ e. Where are the multiples of 6? How are the primes related to the multiples of 6?

   f. Where are the twin primes? What can you say about the number between any two twin primes?

★ g. The number 12 has a prime before it and a prime after it. Does every multiple of 12 have one or two adjacent primes?

5. Each time the grid is changed we can expect different patterns of multiples and primes.** The grid shown here is a counterclockwise spiral of numbers beginning with 1 in the center. Circle the primes in this grid.

★ a. Examine the diagonals from lower left to upper right. Which of these diagonals contain: even numbers? odd numbers? prime numbers?

   b. Answer the questions in part **a**, for the diagonals from lower right to upper left.

*K.P. Swallow, "The Factorgram," *The Mathematics Teacher,* 48 (January 1955), pp. 13–17.
**F. Tapson, "Mining for Numbers," *Mathematics in Schools,* 2 No. 6 (November 1973), pp. 2–4.

c. Which diagonals do not contain prime numbers, other than the number 2?

★ d. What is the greatest number of primes in consecutive squares along the diagonals of this grid?

6. A similar spiral grid with whole numbers from 1 to 10,000 was used by Stanislaw M. Ulam of the Los Alamos Scientific Labs to look for patterns of primes. The locations of these primes are pictured in this photograph of a computer grid.

   a. Explain why the primes cluster along the diagonals, rather than by rows or columns.

   b. What appears to be the greatest number of primes in consecutive positions along the diagonals of this grid?

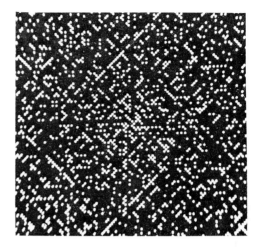

Just for Fun – Spirolaterals

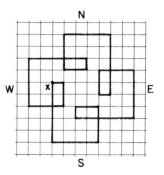

Spirolaterals are graphs of number sequences. The spirolateral shown in the accompanying diagram is the graph of 1,2,3,4,5. Starting at point *x*, the following segments were drawn: a segment of length 1 in the East direction; a segment of length 2 in the South direction; a segment of length 3 in the West direction; and so on, repeating the number sequence with a 90-degree clockwise turn each time. After the sequence 1,2,3,4,5 is repeated several times the spirolateral will be *completed*, in that it will have returned to its starting point and further use of the sequence will result in a retracing of the same figure. How many times was the sequence 1,2,3,4,5 used before the spirolateral was completed?

The number of numbers in a sequence is called the *order of the spirolateral.* Some spirolaterals cannot be completed but tend to wander off in some direction. Whether or not this happens depends on the order of the spirolateral. Sketch the spirolaterals
★ shown next, beginning at the given starting points. Which two of these spirolaterals cannot be completed? What happens to the spirolateral if the numbers in its sequence are reversed? (For more graph paper, see Material Card 14.)

a. 1,2,3   b. 1,2,3,4   c. 1,2,3,4,5,6   d. 1,3,3,1,5,2   e. 1,2,3,4,5,6,7,3

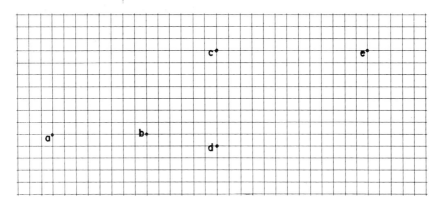

In order for a spirolateral to be completed, the total movement in the East direction must equal the total movement in the West direction. Similarly, there must be a balance of total movement in the North and South directions. The amount of movement in each of these four directions, for the spirolateral 2,5,1,3,3,1, is shown in the table on the right. The sums of numbers in the East-West directions are 6 and the sums in the North-South directions are 9. This spirolateral will close after the numbers are used twice. Try it. Complete the following E S W N tables for the spirolaterals which were graphed above. Two of these tables cannot be completed so that the sums in the columns balance.

E	S	W	N
2	5	1	3
3	1	2	5
1	3	3	1
6	9	6	9

★ a. E S W N   ★ b. E S W N   ★ c. E S W N   d. E S W N   e. E S W N

The only spirolaterals in the preceding examples which could not be completed were the two whose orders were multiples of 4. This suggests the following conjecture: If the order of a spirolateral is not a multiple of 4, the spirolateral can be completed. Test this statement for some more spirolaterals. Try your telephone number or your zip code.

At this point, you may be tempted to conclude that if the order of a spirolateral is a multiple of 4, the spirolateral cannot be completed. Before you do, try graphing the following spirolaterals:

1,2,1,2;   1,2,3,4,4,3,2,1;   and   1,2,4,4,4,5,1,3

The spirolateral in the opening example, 1,2,3,4,5, was completed by using the numbers in its sequence four times. Try a few more spirolaterals to test the following conjecture: If the order of a spirolateral is odd, the number sequence will be needed four times before the spirolateral is completed. *Note:* Spirolateral 1,2,3,4,8,2,6 will return to its starting point after the sequence has been used only once. However, the spirolateral will not be complete at this point, because a continuation of the numbers does not give a retracing of the spirolateral.)

Graph a few spirolaterals whose orders are even numbers but not multiples of 4. Form a conjecture about the number of times their sequences are needed before the spirolaterals are completed.

★ If a spirolateral can be completed, its sequence of numbers will be used one, two, or four times. Use the concept of least common multiple to explain why.

Spirolaterals can also be graphed on isometric grids (Material Card 15). The graph of 1,2,3,4 is shown here. The starting point is x and each turn is $120°$ clockwise. Graph a few spirolaterals on your isometric grid and form some conjectures about which ones can be completed. There are three directions on an isometric grid: East (E), Southwest (SW), and Northwest (NW). Try analyzing these spirolaterals by using an E SW NW table.*

ACTIVITY SET 3.6

FACTORS AND MULTIPLES WITH CUISENAIRE RODS

"Mathematics is the queen of the sciences but number theory is the queen of mathematics."

Carl Friedrich Gauss

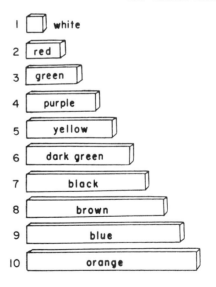

Greatest common factor and least common multiple are important concepts in number theory. A pair of numbers may have several common factors. For example, 1, 2, 3, and 6 are all factors of both 12 and 18. The *greatest common factor* of 12 and 18 is 6: g.c.f. (12,18) = 6. There are an infinite number of multiples of both 12 and 18. The first few are 36, 72, 108, and 144. The *least common multiple* of 12 and 18 is 36: l.c.m. (12,18) = 36. In the following activities, factors, multiples, greatest common factors, least common multiples, and prime and composite numbers will be illustrated by Cuisenaire rods. These rods will be used to represent whole numbers from 1 to 10; however, they can

---

*Additional information about spirolaterals can be found in F. Odds, "Spirolaterals," *The Mathematics Teacher,* **66** No. 2 (February 1973), pp. 121-24.

60 CHAP. 3 WHOLE NUMBERS AND THEIR OPERATIONS

be assigned different values (see Activity Set 6.1). Numbers greater than 10 are represented by placing rods end to end in a linear figure which is called a *train*. There are copies of these rods on Material Card 37.

1. This multicolored train of rods and the single-color train both have the same length and each represents the same number.

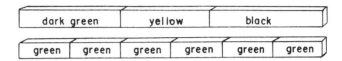

★ a. The single-color train shows two of the factors of the number represented by the multicolored train. What are these factors?

b. Build as many different single-color trains as you can whose length equals that of the multicolored train. How many trains are there in all? How many factors are there of the number represented by the multicolored train?

★ c. In a single-color train the length of each rod is one factor of the number represented by the train, and the number of rods is another factor. Does this mean that every number has an even number of factors?

2. Form a multicolored train by joining rods having the following colors: white, yellow, white, purple, white, red, green, dark green, white, green, and red.

a. Can you form an all-red train whose length equals the length of this train?

b. Can you form an all-green train whose length equals the length of this train?

★ c. What conclusions can be drawn from parts **a** and **b** regarding the number represented by the original train?

3. Form the smallest all-yellow train and the smallest all-green train that are both equal in length.

★ a. List several number facts which these rods and trains illustrate.

b. Does this activity give information about a greatest common factor or a least common multiple? What is this information?

4. Try forming single-color trains, using the same color for both, which are equal to the lengths of the following two trains.

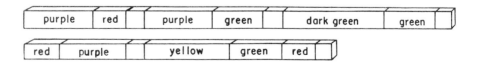

ACTIVITY SET 3.6

a. Can this be done with green rods?

b. Can two such trains be formed with yellow rods?

c. What is the biggest rod that can be used to form two such trains?

★ d. What do the three single-color trains in parts **a**, **b**, and **c** indicate about the two given trains shown previously? List several facts.

e. Do these trains give information about greatest common factor or least common multiple? What is this information?

5. The product of two numbers is represented by crossing the two rods for these numbers, as shown in Figure 1. The green and yellow rods represent 3 times 5. Rods crossed in this manner are called a *tower of rods*. This arrangement of crossed rods is used to represent 3 × 5 because if we form a rectangle which has the length of a yellow rod and the width of a green rod (see Figure 2), it will cover the same region as 15 white rods.

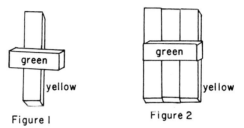

Figure 1   Figure 2

Use your rods to form a product tower for each of the following numbers. List the numbers of each type of rod in your towers.

★ a. 72   b. 63   c. 105

6. This tower of purple, green, and red rods is one of the towers for 24. It represents 4 × 3 × 2.

   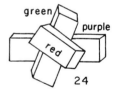

   a. There are four more product towers with different combinations of the ten types of rods which can be formed for 24. List the numbers of each type of rod for these towers.

★ b. The tallest tower for 24 has only red and green rods. Explain why any tower of rods can be replaced by a tower which contains only red, green, yellow, or black rods.

7. Represent each of the following numbers by two different towers, so that one of these towers contains only rods for prime factors. List the rods in each tower. For each number, which of the two towers is the taller?

   a. 70   b. 210   c. 168

8. When two numbers are represented by towers of prime factors, their greatest common factor is obtained from the rods that are common to both towers. The green and black rods are common to the towers for 42 and 105, so these rods represent the factors of the greatest common factor of the two numbers. Thus, the greatest common factor is 3 × 7 or 21.

Build towers of prime factors for each of the following pairs of numbers. Mark the types of rods shown in the figures which are common to both towers. Determine the greatest common factor of each pair of numbers.

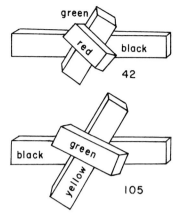

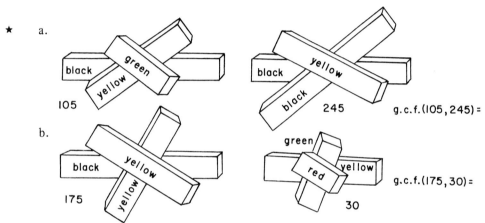

★ a. 105   245   g.c.f.(105, 245) =

b. 175   30   g.c.f.(175, 30) =

9. When two numbers are represented by towers of prime factors, their least common multiple is found by building a tower having the maximum number of different rods contained in the two towers. Build towers of prime factors for each of the following pairs of numbers. For each pair of towers mark the rods that would be used in the tower for the least common multiple.

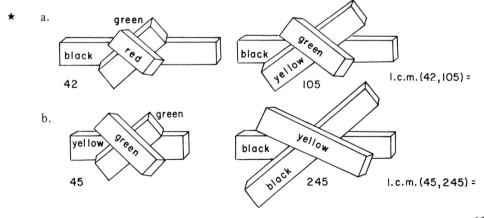

★ a. 42   105   l.c.m.(42, 105) =

b. 45   245   l.c.m.(45, 245) =

ACTIVITY SET 3.6

10. **Challenge:** Using only red, green, yellow, and black rods, build the tower for the smallest number having 1, 2, 3, 4, 5, 6, 7, 8, 9, and 10 as factors. List the numbers of each type of rod in this tower.

Just for Fun—**Star Polygons**

Star polygons are often constructed to provide decorative and artistic patterns. The star polygon pictured to the right was formed by colored string and 16 equally spaced tacks on a piece of plywood. In the following activities these patterns are analyzed by using the concepts of factor, multiple, greatest common factor, and least common multiple.*

Star polygons can be constructed by taking steps of a given size around a circle of points. Star (14,3) was constructed by beginning at point p and taking a step of 3 spaces to point q. Three spaces from q is point r. This process will eventually bring us back to point p, after having "hit" all 14 points. The resulting figure is a star polygon.

In general, for whole numbers n and s, star (n,s) will denote a star polygon with n points and steps of s, provided that $s < n$. In the special cases where $s = 1$ or $s = n - 1$, the resulting figure will be a polygon.

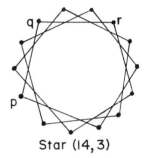

Star (14, 3)

★ 1. Sketch the following star polygons by beginning at point x and taking steps of s in a clockwise direction. Will the same star polygons be obtained if the steps are taken in a counterclockwise direction?

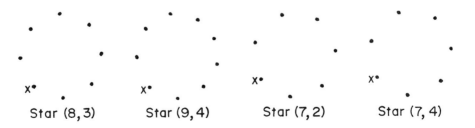

Star (8, 3)   Star (9, 4)   Star (7, 2)   Star (7, 4)

*A.B. Bennett, Jr., "Star Patterns," *The Arithmetic Teacher,* 25 No. 4 (January 1978), pp. 12–14.

★ 2. Sketch the following pairs of star polygons. Make a conjecture about star (n,s) and star (n,r) where $r + s = n$.

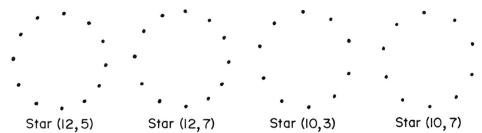

Star (12, 5)   Star (12, 7)   Star (10, 3)   Star (10, 7)

★ 3. The star polygons in Activities 1 and 2 can each be completed by beginning at any point and drawing one continuous path. In each of these examples the path returns to the starting point after "hitting" all points. For star (15,3), however, the path closes after "hitting" only 5 points. To complete this star, 3 different paths will be needed. Determine the number of different paths for each of the following star polygons.

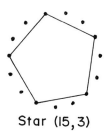

Star (15, 3)

Star (5, 1)   Star (10, 4)   Star (12, 4)   Star (6, 3)

★ 4. What conditions must n and s satisfy in order for star (n,s) to be formed with 1 continuous path? Use your conjecture to sketch the star polygons having 15 points which can be formed by 1 continuous path. Star (15,1) and star (15,14) is one of these. Sketch the remaining three star polygons and write the number of steps for each.

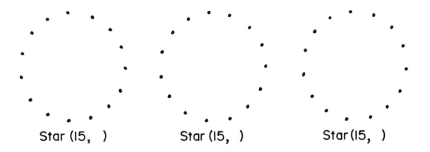

Star (15,)   Star (15,)   Star (15,)

ACTIVITY SET 3.6

★ 5. Star (14,3) was constructed by beginning at point *p* and taking steps of 3 to points *q, r, s, t*, etc. After point *t* the next step of 3 completes 1 orbit of the circle (once around the circle) and starts us on the second orbit. It takes 3 orbits before the path returns to its starting point. How many orbits are needed for the following star polygons before a given path will return to its starting point?

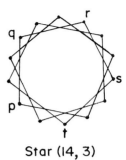

Star (14, 3)

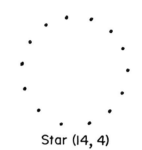

Star (9, 2)   Star (7, 3)   Star (14, 4)

6. The star (7,3) has been drawn by starting at point 1 and connecting the points in the following order: 1, 4, 7, 3, 6, 2, 5, and 1. Three orbits are needed to complete this star. The number of orbits is just the number of times we need the 7 points of the circle before the steps of 3 bring us back to the beginning point. That is, after taking steps of 3 for 21 spaces we arrive back at the beginning point. The 21 spaces is the least common multiple of 7 and 3.

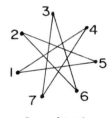

Star (7, 3)

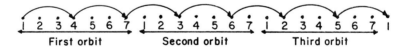

a. Explain how the concept of least common multiple can be used to determine the number of orbits for 1 path of star (15,6). Check your answer.

★ b. Make a conjecture about the number of orbits to complete 1 path for star (n,s).

7. Create a star polygon. Any number of points may be used and they do not have to lie on a circle. Star polygons can be made by drawing with colored pencils (different colors for steps of different sizes) or by using colored thread or yarn. With a needle and thread, they can be stitched on posterboard. Several star polygons can be formed on the same set of points. Three star polygons have been stitched from 52 holes on 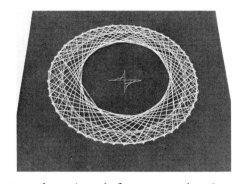 the black posterboard pictured here. One star polygon is made from orange thread with steps of 2; another from green thread with steps of 10; and one from yellow thread with steps of 15.

   a. The star polygon with steps of 15 is the only one which can be constructed by 1 continuous path. Why?

★ b. How many orbits are needed to complete star (52,15)?

4

Geometric Figures

ACTIVITY SET 4.1

RECTANGULAR AND CIRCULAR GEOBOARDS

"That mathematics is the handmaiden of science is a commonplace; but it is less well understood that experiments stimulate mathematical imagination, aid in the formation of concepts, and shape the direction and emphasis of mathematical studies. One of the most remarkable features of the relationship is the successful use of physical models and experiments to solve problems arising in mathematics."

James R. Newman

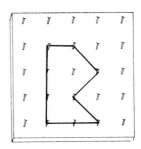

This quote by Newman seems an appropriate introduction to the geoboard, which for more than a decade has been a popular physical model for illustrating geometric terms and properties. The most familiar type of geoboard has a square shape with 25 nails arranged in a 5 by 5 array. By placing rubber bands on the nails, models for line segments, angles, and polygons can be formed. In particular, the rectangular geoboard will be used in the following activities to form: convex and nonconvex polygons; isosceles, right, and scalene triangles; and rectangles, parallelograms, and trapezoids. The polygon on this geoboard is nonconvex. One test for nonconvexity is to connect boundary points with a band. If there are two boundary points for which the band between them falls outside the polygon, the polygon is *nonconvex*.

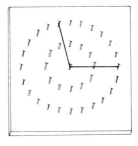

The circular geoboard is more recent than the rectangular (square) geoboard, and has not been used as extensively. The geoboard shown here has a nail at the center, an inner circle of 12 nails, and an outer circle of 24 nails. Many of the angles and polygons which can be formed on this geoboard are not possible on the rectangular geoboard. The circular geoboard will be used to form central and inscribed angles and to show an important relationship between them.

Since most geoboard activities can be carried out on dot paper by drawing figures, the use of geoboards is optional in the following activities. However, geoboards provide an added stimulation, and the ease with which figures can be shaped encourages experimentation and creativity. If geoboards are not available you may wish to make them from boards and finish nails. There are rectangular and circular geoboard templates on Material Cards 17 and 18 for spacing the nails.

ACTIVITY SET 4.1   69

1. Which one of these hexagons is non-convex?

★ a. A *quadrilateral* is a polygon with four sides. Squares, rectangles, and parallelograms are convex quadrilaterals. Can a nonconvex quadrilateral be formed on the geoboard?

b. Form the following polygons on your geoboard and sketch them here. (A pentagon has 5 sides; a heptagon has 7 sides; and an octagon has 8 sides.)

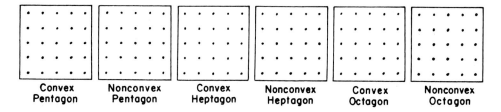

Convex Pentagon   Nonconvex Pentagon   Convex Heptagon   Nonconvex Heptagon   Convex Octagon   Nonconvex Octagon

★ c. What is the convex polygon with the greatest number of sides which can be formed on the geoboard?

2. There are many geoboard activities in which the object is to form as many polygons as possible of a given type. Here are a few examples.

   a. A triangle is *isosceles* if two of its sides have the same length. There are 8 noncongruent isosceles triangles which can be formed with their bases parallel to the lower edge of the geoboard. Sketch two on each geoboard.

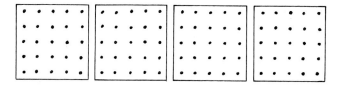

★ b. There are 8 noncongruent triangles which can be formed on a 3 by 3 geoboard. Sketch them here. A triangle with a right (90°) angle is a *right triangle*, and if all three sides have different lengths the triangle is called *scalene*. Which of these triangles are: isosceles? right? scalene?

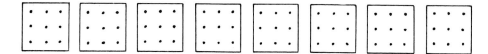

70   CHAP. 4   GEOMETRIC FIGURES

★ c. There are 16 noncongruent quadrilaterals which can be formed on a 3 by 3 geoboard. Sketch them here. Which of these quadrilaterals are: squares? rectangles but not squares? parallelograms but not rectangles? trapezoids but not parallelograms? nonconvex? (A parallelogram has two pairs of opposite parallel sides, and a trapezoid has one pair of opposite parallel sides.)

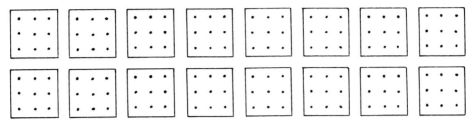

3. The intersection of these two rectangular regions is a triangular region and their union is a heptagonal region. Try to form the polygons whose regions satisfy the following conditions. Sketch your answers on the geoboards that follow. One of these cannot be done.

★ a. Two rectangles whose intersection is a square and whose union is a hexagon.

b. Two rectangles whose union is a triangle.

c. A quadrilateral and a pentagon whose union is an octagon.

★ d. Three triangles whose union is a rectangle.

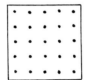

★ 4. **Mystery Polygon**: Use the following clues to form the polygon which is described. Sketch it here.

a. This polygon touches only 6 nails.

b. It is nonconvex.

c. It has no interior nails (nails which do not touch the boundary).

d. Five of its sides are parallel to the sides of the geoboard and have the same length.

ACTIVITY SET 4.1 71

★ 5. **Central Angles:** An angle whose vertex is the center of a circle and whose sides intersect the circle is called a *central angle*. The portion of the circle which is cut off by the sides of the angle is called the *intercepted arc*. The central angle on this geoboard has a measure of 15° because its intercepted arc is 1/24 of a whole circle.*

★   a. Under each geoboard write the number of degrees in the central angle.

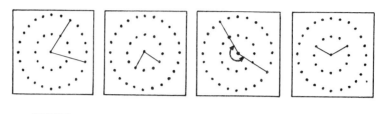

★   b. Form central angles on your geoboard which have the following numbers of degrees.

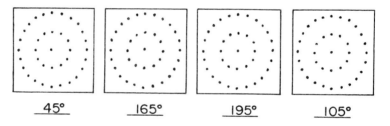

       45°   165°   195°   105°

★ 6. **Inscribed Angles:** An angle is called an *inscribed angle* if its vertex is on a circle and its sides intercept an arc of the circle. Angle *ABC* is an inscribed angle.

★   a. The size of an inscribed angle depends on the size of its intercepted arc. For example, angle *RST* and angle *HIJ* are congruent because their intercepted arcs are congruent. Sketch another angle on these geoboards which is congruent to the given angles.

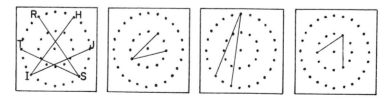

*S. Jencks and D. Peck, "Thought Starters for the Circular Geoboard," *The Mathematics Teacher,* **67** No. 3 (March 1974), pp. 228–33.

★ b. Angle *PQR* is inscribed in a semicircle (half of a circle). On each of the other geoboards sketch an angle which is inscribed in the upper semicircle and whose sides intersect the ends of the diameter that has been drawn. What is the number of degrees in each of these angles?

★ 7. The inscribed angle, angle *ABC*, and the central angle, angle *AOC*, both intercept the same arc. How many degrees are in the central angle? Use your protractor (Material Card 41) to find the number of degrees in the inscribed angle.

Angle AOC ____
Angle ABC ____

★ a. On each of the following geoboards draw the central angle which intercepts the same arc as the given inscribed angle. Write the number of degrees in each angle.

Inscribed Angle____   Inscribed Angle____   Inscribed Angle____   Inscribed Angle____
Central Angle____   Central Angle____   Central Angle____   Central Angle____

★ b. There is a way to determine the number of degrees in an inscribed angle by knowing the number of degrees in the corresponding central angle. How can this be done?

8. **Mystery Polygon:** Use the following clues to form the polygon which is described. Sketch it here.
   a. The polygon has less than 5 sides.
   b. The polygon is not convex.
   c. Two of the sides of the polygon are radii of the outer circle.
   d. The 2 radii form a right angle.
   e. This polygon touches only 6 nails.
   f. There is 1 nail inside the polygon.

ACTIVITY SET 4.1 73

Just for Fun—Tangram Puzzles

Tangrams are some of the oldest and most popular of the ancient Chinese puzzles. The seven tangram pieces are shown in this square. There are 5 triangles, 1 square, and 1 rhombus. Fold a standard piece of paper to get a square and cut it out. Then by paper folding obtain the seven tangram pieces and cut them out for the following activities.

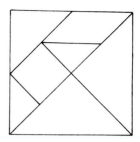

These seven pieces are used in Chinese puzzles called tangram puzzles. The object is to use all of the tangram pieces to form figures. The tangram pieces should be placed so that they are nonoverlapping. Try forming the following figures.

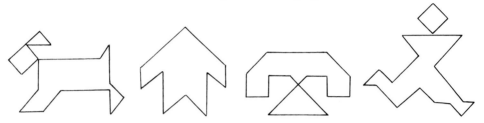

It is possible to form several squares using different combinations of tangram pieces. Squares can be made from 1, 2, 3, 4, 5, and 7 tangram pieces. Form these squares and sketch the solutions.

It is possible to form several triangles and parallelograms with different combinations of tangram pieces. The triangles can be made from 1, 2, 3, 4, 5, or 7 tangram pieces and parallelograms with 1, 2, 3, 4, 5, 6, or 7 pieces. Form these figures and sketch the solutions.

There are many tangram figures. You may wish to try making a few of your own. The book *Tangrams* by Ronald Read contains 330 figures and their solutions.*

*R. Reed, *Tangrams* (New York: Dover, 1965).

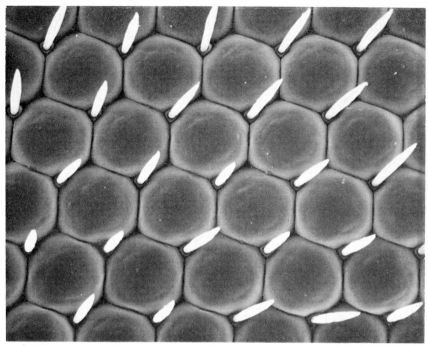

A portion of the eye of a fruit fly

ACTIVITY SET 4.2

REGULAR AND SEMIREGULAR TESSELLATIONS

The hexagonal shapes in the above picture cover a small portion (about one-tenth) of the eye of a fruit fly. The partitioning of surfaces by hexagons, or nearly hexagonal figures, is common in Nature.* A figure such as a regular hexagon, which can be used repeatedly to cover a surface without gaps or overlaps, is said to *tile* or *tessellate* the surface. The resulting pattern is called a *tessellation.* The basic figure for a tessellation can have an infinite variety of shapes. The famous Dutch artist Maurits C. Escher (1898-1972) is noted for his tessellations with drawings of birds, fish, and other living creatures. A technique for creating Escher-type tessellations follows this activity set.

*For some examples of hexagonal patterns and a discussion of why they occur in Nature, see H. Weyl, *Symmetry* (Princeton: Princeton University Press, 1952), pp. 83-89.

1. This hexagonal skylight is a tessellation of equilateral triangles. Each vertex point of this tessellation is surrounded by six triangles.

    a. How many degrees are there in each angle of an equilateral triangle? Explain why equilateral triangles tessellate.

    ★ b. The six triangles meeting at the center of this picture form a regular hexagon. This hexagon is surrounded by six more hexagons. How many degrees are there in each angle of a regular hexagon? Explain why regular hexagons tessellate.

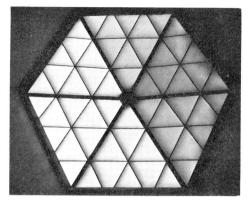

Skylight in New England Center, Durham, N.H.

2. Here are the first six regular polygons. For each polygon trace additional copies around the enlarged vertex point, as if to begin a tessellation.

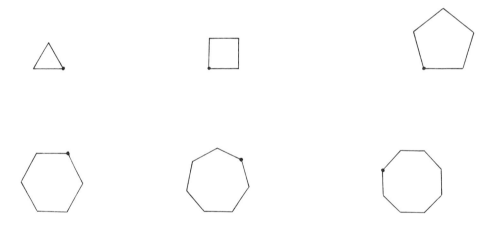

    a. Which of these polygons will not tessellate?
    ★ b. What condition must the angles of a regular polygon satisfy in order for the polygon to tessellate?
    ★ c. Explain why regular polygons with more than six sides will not tessellate.

76 CHAP. 4 GEOMETRIC FIGURES

3. Some convex pentagons will tessellate because their angles have the right numbers of degrees and their sides have matching lengths.* In the pentagon at the right, angle *BAE* has a measure of 60° and angle *CDE* has a measure of 120°.

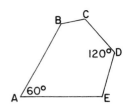

   a. Name two pairs of sides which have the same lengths. In a tessellation of this pentagon, pairs of sides with the same lengths will be matched up. Form a portion of this tessellation.

   b. How many angles meet at vertex *A* for this tessellation? What is the number of degrees in each of these angles?

★ c. How many angles meet at vertex *D* for this tessellation? What is the number of degrees in each of these angles?

★ 4. You have seen that regular hexagons will tessellate. Engineers at IBM have used this fact in developing a system for storing computer information which uses data-filled cartridges in hexagonal compartments. There are also convex nonregular hexagons which tessellate. Form a portion of the tessellation for the hexagon at the top of the next page by finding combinations of angles whose measures add up to 360° and by matching sides of equal lengths. (The answers to parts **a** and **b** will help to form this tessellation.)

IBM 3850 Mass Storage System

   a. Name two pairs of sides which have the same lengths. These pairs of sides will be matched up in the tessellation.

★ b. There are two types of vertex points for this tessellation: Three angles whose sum is 360° meet at one type of vertex; and the other three angles whose sum is also 360° meet at the second type of vertex. List the angles at each type of vertex.

*There are eight types of convex pentagons which will tessellate. See M. Gardner, "On tessellating the plane with convex polygon tiles," *Scientific American,* **233** No. 1 (July 1975), pp. 112–17.

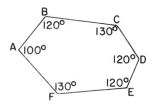

★ 5. The equilateral triangle, square, and regular hexagon are the only regular polygons that tessellate. Although designs based on these patterns are common, more interesting tessellations can be formed by using two or three regular polygons. A square and two octagons are around each vertex point of the tessellation shown here. When there are two or more types of regular polygons whose vertices meet only at other vertices, and for which every vertex is surrounded by the same arrangement of polygons, the tessellation is called *semiregular*. The five regular polygons which are used in semiregular tessellations are shown next. The number of degrees in each vertex angle of each polygon is given above the polygon. There are several copies of each of these polygons on Material Card 23. Cut them out for use with the following activities.

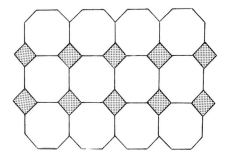

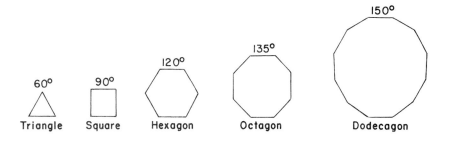

★ a. There are six semiregular tessellations that can be formed using only two of these polygons. Sketch one of these tessellations. (*Hint:* Look for combinations of angles whose measures add up to 360°.)

★ b. There are two semiregular tessellations which use combinations of three of the polygons. Sketch one of these tessellations. (*Hint:* Look for combinations of angles whose measures add up to 360°.)

★ c. There is a larger collection of regular polygons from which we can choose, if we merely wish to place combinations of polygons about a single point so that there are no gaps or overlapping.* For example, an equilateral triangle, a regular octagon, and a regular 24-sided polygon can be placed about a point, because the sum of the degrees from three angles of these polygons is 360°. However, these polygons cannot be used to form a semiregular tessellation. How many degrees are in each vertex angle of these polygons?

6. Once we allow nonconvex polygons for a tessellation there is no limit to the possibilities. Select one of the following polygons and form a portion of a tessellation.

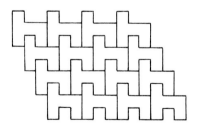

*For a complete list of 17 combinations of such polygons, see D. R. Duncan and B. H. Litwiller, "A Simple Sorting Sequence," *The Mathematics Teacher,* **67** No. 4 (April 1974), pp. 311-15.

Just for Fun — Escher-type Tessellations

This tessellation of birds and fish is a 1952 woodcut by M.C. Escher. Escher was fascinated by the fact that a surface can be filled up with figures without leaving any open space. As he once said, "This is the richest source of inspiration I have ever struck; nor has it yet dried up."

Two Intersecting Planes, by M.C. Escher

The following examples show a simplified step-by-step creation of figures for tessellating. By beginning with polygons which we know will tessellate, such as triangles, squares, rectangles, parallelograms, and so on, modifications can be made to produce a variety of shapes. These shapes may then suggest living figures which can be sketched as shown in the following examples. Create your own pattern and tessellate with it.*

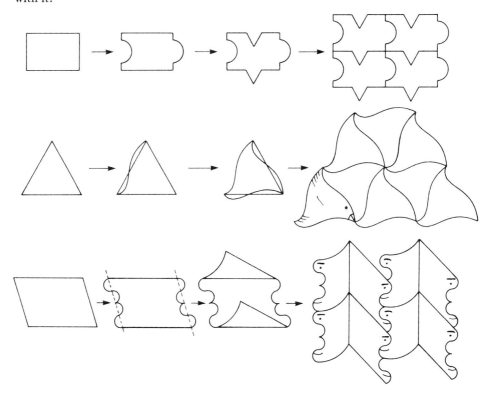

*See J. Teeters, "How To Draw Tessellations of the Escher Type," *The Mathematics Teacher,* **67** No. 4 (April 1974), pp. 307–10.

Stars, by M.C. Escher

ACTIVITY SET 4.3

MODELS FOR REGULAR AND SEMIREGULAR POLYHEDRA

A *regular polyhedron* is a convex polyhedron whose faces are congruent regular polygons and whose vertices are each surrounded by the same number of these polygons. The proof that there are only five such polyhedra is elementary and was known by the early Greeks.*
These polyhedra are called *Platonic solids.* They are named for the Greek philosopher Plato (430 B.C.), who discovered them independently, although four of them were known before his time by the Egyptians. Like Leonardo da Vinci, M.C. Escher had a strong appreciation for the Platonic solids, which can be seen in his wood engraving, *Stars.*

---

*For a proof that there are exactly five regular polyhedra, see P.G. O'Daffer and S.R. Clemens, *Geometry: An Investigative Approach* (Reading, Massachusetts: Addison-Wesley, 1976), pp. 115–19.

1. **Platonic Solids:** There are several common methods of constructing the Platonic solids. The polyhedra pictured below are "see through" models, similar to those in Escher's wood engraving. They are made from drinking straws which can be threaded or stuck together. Another method is provided by the patterns on Material Cards 38 and 39. Cut these out and use a ball-point pen or sharp point to score the dotted lines. Fold on these lines and tape the edges to form the Platonic solids.

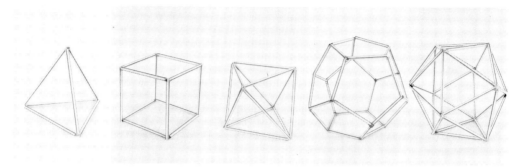

★ a. Each Platonic solid is named according to its number of faces. How many faces does each polyhedron have, and what regular polygon is used for each face?

b. The faces of the icosahedron and cube in this photo were made from Christmas cards. The triangular and square patterns for these faces are shown below. The corners of these patterns were cut with a paper-punch and scissors, and the edges bent up along the dotted lines. The adjoining faces of these polyhedra are held together by rubber bands on their edges. How many rubber bands were needed to make each of these polyhedra (one band was used for each matching pair of edges)?

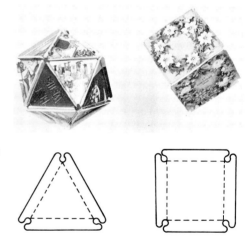

c. You may enjoy planning colors for for your Platonic solids, so that no two faces with a common edge have the same color.* For this condition what is the least number of colors needed for the octahedron?

*For some suggested color schemes, see M.J. Wenninger, *Polyhedron Models* (Reston, Virginia: National Council of Teachers of Mathematics, 1966).

★ 2. **Euler's Formula:** There is a simple formula which relates the numbers of vertices (v), faces (f), and edges (e), of a polyhedra. The cube has 8 vertices, 6 faces, and 12 edges. In the table on the right, list the numbers of the vertices, faces, and edges for the remaining polyhedra. Look for a relationship between these numbers, for each polyhedron.

	Vertices	Faces	Edges
Tetrahedron			
Cube	8	6	12
Octahedron			
Dodecahedron			
Icosahedron			

3. **Patterns for Platonic Solids:** Patterns (1), (2), and (3) were formed by joining 6 squares along their edges. Pattern (1) occurs on Material Card 38 and can be folded into a cube.

★ a. Which one of the two patterns, (2) or (3), can be folded into a cube?

b. There are 11 different patterns of 6 squares which can be folded into a cube. Sketch another one.*

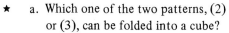

★ c. There are also several patterns of triangles which can be folded into an octahedron. Which one of the two patterns, (4) or (5), can be used? Make a sketch of another pattern which can be folded into an octahedron. (*Hint:* Make an octahedron and then cut it apart to obtain a pattern.)

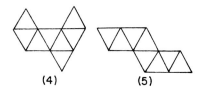

d. Pattern (6) was used on Material Card 38 for making a tetrahedron. Sketch another pattern which can be folded into a tetrahedron.

---

*One way of obtaining these patterns is to begin with a cube, such as a milk carton, and work backwards to a pattern. For more information on these patterns, see E.R. Ranucci and W.E. Rollins, *Curiosities of the Cube* (New York: Thomas Y. Crowell, 1977), pp. 6–7.

4. **Archimedean Solids:** A *semiregular polyhedron* is a polyhedron which has two or more regular polygons as faces and for which each vertex is surrounded by the same arrangement of polygons. The 13 semiregular polyhedra are shown below. They are also called *Archimedean solids.* Archimedes (287–212 B.C.) wrote a book on these solids but it has been lost. The six polyhedra, **b, e, f, h, j,** and **l,** can be obtained by cutting off the corners of regular polyhedra. For example, if each corner of the tetrahedron is cut off we obtain the truncated tetrahedron having four hexagons and four triangles as faces (**l**). Find the polyhedra that **b, e, f, h,** and **j** can be obtained from. The remaining seven polyhedra can be obtained by further cuts on the corners of polyhedra **b, e, f, h, j,** and **l.** The 13 Archimedean solids can be constructed by using the patterns on Material Card 24 for their faces.

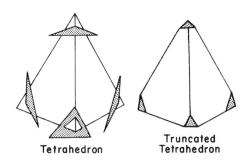

Tetrahedron   Truncated Tetrahedron

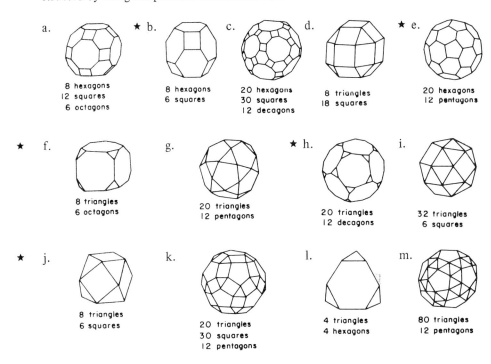

a. 8 hexagons / 12 squares / 6 octagons

★ b. 8 hexagons / 6 squares

c. 20 hexagons / 30 squares / 12 decagons

d. 8 triangles / 18 squares

★ e. 20 hexagons / 12 pentagons

★ f. 8 triangles / 6 octagons

g. 20 triangles / 12 pentagons

★ h. 20 triangles / 12 decagons

i. 32 triangles / 6 squares

★ j. 8 triangles / 6 squares

k. 20 triangles / 30 squares / 12 pentagons

l. 4 triangles / 4 hexagons

m. 80 triangles / 12 pentagons

84 CHAP. 4 GEOMETRIC FIGURES

5. **Deltahedra**: This figure is convex and has 10 equilateral triangles for faces. There are exactly eight convex polyhedra whose faces are equilateral triangles. They are called *deltahedra,* after the Greek letter △.

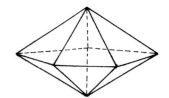

★ a. Explain why this 10-sided deltahedron is not a regular polyhedron. (*Note:* There are five triangular faces meeting at the top and bottom vertices.)

  b. Three of the deltahedra are Platonic solids. What are their names?

★ c. Each two faces of the 10-sided deltahedron shares a common edge. Therefore, the total number of edges is 10 × 3 divided by 2. A deltahedron with n triangular faces will have $3n/2$ edges. Explain why $3n/2$ implies that n, the number of faces in a deltahedron, must be even.

  d. The eight deltahedra have 4, 6, 8, 10, 12, 14, 16, and 20 faces. Cut out sections of the isometric grid on Material Card 40 to make one of the deltahedra which has either 6, 12, 14, or 16 faces. Explain why the deltahedron you make is not a regular polyhedron.*

Just for Fun—Penetrated Tetrahedron

The penetrated tetrahedron consists of two tetrahedra, one of which is inside the other and penetrating its sides. To have your own penetrated tetrahedron follow the steps listed next.

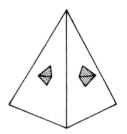

Step 1: Draw equilateral triangle ABC. Draw the medians from the three vertices to the midpoints of the opposite sides. Label the intersection of the medians by O and the midpoint of line segment $\overline{AB}$ as D.

Step 2: Locate point K on $\overline{CD}$ so that AK is equal to CD. Mark points X, Y, and Z on the medians so that their distances from O are equal to CK. Form triangle XYZ and cut it out.

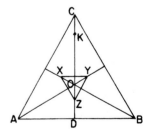

---

*William E. McGowan, "A Recursive Approach to the Construction of the Deltahedra," *The Mathematics Teacher,* **71** No. 3 (March 1978), pp. 204-10.

Step 3: Use some sturdy material (oaktag or construction paper) and form a pattern with 4 copies of triangle *ABC* as shown here. Cut out the 4 smaller triangles from the center of each large triangle.

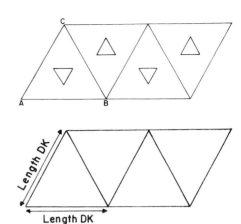

Step 4: Make another pattern of 4 equilateral triangles with sides of length *DK*. Assemble this tetrahedron and insert it into the larger tetrahedron which can be formed by the pattern in Step 3.

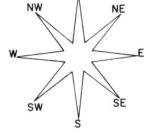

ACTIVITY SET 4.4

CREATING SYMMETRIC FIGURES BY PAPER FOLDING

The *wind rose* is a mariner's device for charting the directions of the wind. The earliest known wind rose appeared on the ancient sailing charts of the Mediterranean pilots who charted eight principal winds. These are marked on the 8-pointed wind rose shown here. Later, half-winds led to a 16-pointed wind rose and quarter-winds brought the total number of points to 32.*

The 8-pointed wind rose is highly symmetric. It has 8 lines of symmetry and 8 rotational symmetries. For example, a line from any two opposite points of the wind rose is a *line of symmetry*. Rotations of 45°, 90°, 135°, etc., are *rotational symmetries*. There are several basic paper-folding patterns in the following activities for cutting out symmetric figures. The variety of figures which can be obtained from slight changes in the angles of the cuts is surprising. The 8-pointed wind rose can be cut from one of these patterns. Try to predict which one, as you do the folding and cutting.

1. Fold a rectangular piece of paper in half twice, making the second fold perpendicular to the first. Let *C* be the corner of all the folded edges. Make two cuts from the edges to an inside point. Predict what kind of polygon you will get before opening the paper. Is the figure convex? How many lines of symmetry does it have?

*M.C. Krause, "Wind rose, the beautiful circle," *The Arithmetic Teacher,* **20** No. 5 (May 1973), p. 375.

★ a. Find a way to make two cuts into an inside point so that you get: a *regular octagon;* a *regular hexagon.* Sketch the location of your cuts.

Octagon

Hexagon

b. Make two cuts off the corner of the folded pattern. Are there any lines of symmetry for the piece between the cuts? Vary the angle of the cuts. Can you get a rhombus inside a square? A square inside a rhombus?

2. The following pattern leads to a variety of symmetrical shapes. Begin with a standard sheet of paper as in Figure (a) and fold down the upper-right corner to produce (b). Then fold the upper vertex R down to point S to obtain (c). Finally, fold this figure over line L to get (d).

(a)

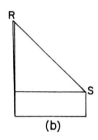
(b)

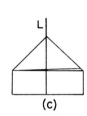
(c)

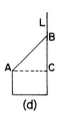
(d)

Perform the indicated cuts in parts **a–h** and draw a picture of the cut-off pieces. Find the number of lines of symmetry for each. Are the crease lines always lines of symmetry? Are there any lines of symmetry which are not crease lines? (Use a separate sheet of paper for each figure.)

★ a.   Cut parallel to line $\overline{AC}$

★ c.   Cut below and parallel to $\overline{DC}$

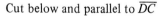

b.   Cut from the midpoint of $\overline{AB}$ to point C

d.   Cut with a greater slope than that of $\overline{DC}$

ACTIVITY SET 4.4

★ e. Cut parallel to $\overline{AB}$

g. Combination of any two cuts

★ f. Cut at an angle to $\overline{CB}$

h. Two cuts into a point inside

3. Figures (a)–(c) in this next folding pattern are the same as those shown in Activity 2. This time, to get (d) fold the two halves inward so that points S and T are on line L.

(a)

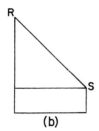
(b)

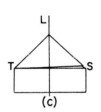
(c)

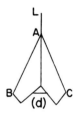
(d)

Try the following cuts on the pattern in (d). Sketch the cut-off pieces and find the number of lines of symmetry for each piece. (Use a separate sheet of paper for each figure.)

★ a. Cut horizontally

c. Cut parallel to $\overline{AB}$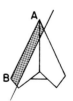

b. Cut slanted

d. Combination of any two cuts

88 CHAP. 4 GEOMETRIC FIGURES

★ e. Two cuts into an inside point on the center line

★ f. Find two cuts into a point on the center line (as in part e) which will produce a regular 16-sided polygon.

★ 4. Many of the cuts on the previous patterns resulted in variations of four-sided and eight-sided polygons. For some different results try some cuts off the pattern shown below in Figure (d).

Figure (a) can be obtained by folding a piece of paper in half twice, with the second fold perpendicular to the first. Point C is the corner of all the folded edges. Draw a line segment down the center of the paper as shown. Next fold the lower-right vertex of the paper up so that it is on the center line as in (b). Now fold the upper left-hand corner down to obtain (c).

Mark points A and B as shown in (d) so that AC = BC. Try to predict what you will get by cutting along $\overline{AB}$. Will the figure be a regular polygon? How many lines of symmetry will there be? Find a way to produce two cuts into this pattern to produce a regular 18-sided polygon.

 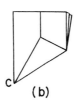 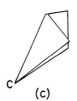
(a)　　　(b)　　　(c)　　　(d)

5. The five-pointed star has been adopted for badges and national symbols for centuries. It appears today on the flags of 41 countries and was once used on the back of a United States four-dollar gold piece.

The four steps pictured in Figures (a)–(d) illustrate a paper-folding approach to making a five-pointed star. To obtain (a), fold a standard sheet of paper perpendicular to its longer side. Next, fold point C over to the midpoint of side $\overline{AD}$, as in (b). To get (c) fold up the corner containing point D. For the final step, fold the right side of Figure (c) over to the left side, covering point D. A cut from X to Y such that XY is greater than YZ will produce a five-pointed star. How many lines of symmetry are there for this star?

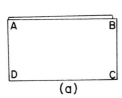   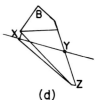
(a)　　　(b)　　　(c)　　　(d)

ACTIVITY SET 4.4

5

Measurement

© 1974 United Feature Syndicate, Inc.

ACTIVITY SET 5.1

MEASURING WITH METRIC UNITS

The basic metric units for length, weight, and volume are the *meter* (a little longer than a yard), the *gram* (about the weight of a paper clip), and the *liter* (a little bigger than a quart). Larger and smaller measures are obtained by using the prefixes listed at the right. Only the four marked with an asterisk are common in everyday use. There are other prefixes for scientific purposes which extend this system to larger and smaller measures. The prefixes differ from one another by multiples of 10. For example,

*kilo . . . . 1000 times
hecto . . . . 100 times
deka . . . . 10 times
*deci . . . . one-tenth
*centi . . . . . one-hundredth
*milli . . . . one-thousandth

10 millimeters (mm) = 1 centimeter (cm);   10 centimeters = 1 decimeter (dm);
10 decimeters = 1 meter (m);   and   1000 meters = 1 kilometer (km).

There is a simple relationship between the measures for length, weight, and volume which makes it easy to convert from one type of measurement to another.

 One cubic centimeter of water equals one milliliter of water and weighs one gram.

The following activities involve the metric units for length and volume. There is a metric ruler on Material Card 41. If you are not familiar with the metric system and are unaccustomed to thinking of length in terms of centimeters and meters, the games at the end of this activity set should be played first.

1. Your handspan is a convenient ruler for measuring lengths. Stretch out your fingers and use your metric ruler to measure the maximum distance from your little finger to your thumb in centimeters.

ACTIVITY SET 5.1   91

a. Use your handspan to approximate the width and length of this page to the nearest centimeter. Then take the measurements with your ruler.

	Width	Length
Approximation		
Measurement		

b. Select some objects around you and estimate their lengths in centimeters. Check your estimations by measuring. Repeat this activity 10 times and see how many distances you can predict to within 1 centimeter. Record the objects, your estimates of lengths, and your measurements.

2. A millimeter is the smallest unit on your ruler. Obtain the measurements of the objects in parts **a** through **c** to the nearest millimeter.
   ★ a. The length of a dollar bill.
   ★ b. The diameter of a penny.
   c. The width of one of your fingernails. Do you have one with a width of 1 centimeter?
   d. What fraction of a millimeter is the thickness of one page of this book? (*Hint:* Measure several pages at once.)

3. This table contains some of the body measurements which are useful for buying and making clothes. Estimate or use your handspan to approximate these measurements. Then use a piece of string and your metric ruler to obtain these measurements to the nearest centimeter.

	Estimate	Measurement
Waist		
Height		
Arm length		
Foot length		
Foot width		
Width of palm		
Hat size		

4. Cut a piece of string so it is 1 meter long.
   ★ a. Use this meter string to measure the distance from the middle of your chest to your fingertips with your arm outstretched. Is this distance greater than or less than 1 meter? This body distance was once used by merchants to measure cloth. It is the origin of the English measure, the yard.
   ★ b. Measure the distance from the floor up to your waist. How does this distance compare with a meter?
   c. Select some objects from around you whose lengths are greater than 1 meter, and try to estimate their length to within 1/2 meter. For example, estimate the height, width, and length of a room and then measure these distances with your meter string. Repeat this activity for 10 different objects and see how many distances you can predict to within 1/2 meter. Record the objects, your estimates of lengths, and the measurements.

5. The kilometer is the metric unit that will replace the mile. Some road signs and automobile speedometers are expressed in both units. A kilometer is about three-fifths of a mile. Use your meter string to measure off 10 meters, and time the seconds it takes you to walk this distance. Use this information to determine how long it would take you to walk 1 kilometer.

6. A liter is very close to the English measure, the quart. A liter is equal to 1000 cubic centimeters, the volume of a 10 cm by 10 cm by 10 cm cube.

★ a. Use the dimensions on this drawing of a quart container to approximately determine the number of cubic centimeters in a quart. Which is greater, a liter or a quart?

   b. What height should be marked on a 1-quart container to measure 1/2 liter?

★ c. What height should be marked on a 2-quart container to measure 1 liter?

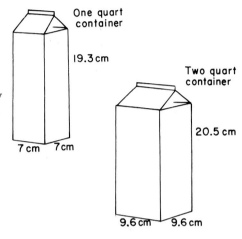

★ 7. **Pendulum Experiment:** There have been many solutions to the problem of measuring time. The shadows cast by sticks, burning candles or rope, hourglasses of sand, and sundials are a few of the ancient methods. Even as late as the time of Shakespeare, people carried pocket sundials. In 1581 Galileo described an application of the pendulum to a clock. The time it takes a pendulum to swing one complete arc is related to the length of the pendulum. Tie a weight on a string and experiment with some pendulums of different lengths. Time their swings and find the length such that one complete swing (over and back) of the pendulum takes 1 second.

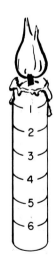

ACTIVITY SET 5.1   93

Just for Fun—Metric Games

Centimeter Racing Game (2 or more players): This game is played on a racing mat (Material Card 20). On a player's turn he/she rolls two dice (or selects numbers from a hat). The two numbers may either be added, subtracted, multiplied, or divided. The resulting computation is the number of centimeters the player moves from the center of the circle towards the triangle. A metric ruler (Material Card 41) should be used to draw one line segment whose length is the player's number. On the player's next turn he/she draws another line segment, beginning at the end of his/her last line segment.

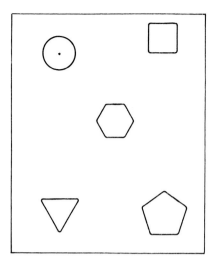

The object of the game is to go from the center of the circle to a point inside the triangle, square, pentagon, and hexagon in that order. Before a player can pass through a polygon or proceed from one polygon to the next, he/she must land inside the polygon with the end of a line segment. The first player to draw a line segment which ends inside the hexagon is the winner.

Centimeter and Meter Guessing Games (2-4 players): On a player's turn he/she places 2 points on a sheet of paper. The next player guesses the distance in centimeters between these 2 points. The distance is then measured with a ruler. If this guess is within 1 centimeter of being correct, the guessing player scores 1 point. If not, he/she receives no score. The players alternate picking points and guessing distances until someone scores 11 points.

Challenging: If you think a player's guess is not within 1 centimeter you may challenge his/her guess by giving your own estimate of the distance. If your estimate is within 1 centimeter and his/hers was not, then you receive 1 point and he/she receives no score. However, if he/she *is* within 1 centimeter, he/she receives 2 points for the insult, regardless of your estimate, and you receive no score.

Variation 1: Play this game by estimating the lengths of objects around you. Since you will be estimating in centimeters, these objects should be reasonably small, such as pencils, books, dollar bills, etc.

Variation 2: This game is called the Meter Guessing Game. On a player's turn, he/she selects an object in the room, and the next player guesses the length in meters. If the guess is within 1/2 meter, the guessing player scores 1 point. If not, he/she receives no score. The players alternate picking objects and guessing distances. As before, a player's guess may be challenged.

ACTIVITY SET 5.2

AREAS ON GEOBOARDS

The rows and columns of nails on the geoboard outline 16 squares. These are the unit squares for the areas of figures. Often these areas can be found by counting the squares and halves of a square inside a figure. This figure contains 6 unit squares and 7 halves of a square, and so its area is 9 1/2 square units. The geoboard will be used in the following activities to examine other methods of finding area and to develop formulas for the areas of triangles, parallelograms, and trapezoids. There is a template for making a rectangular geoboard on Material Card 17.

1. Find the areas of these polygons by counting squares and halves of a square.

   ★ a.   b.   ★ c.

2. In Figure **a** the upper shaded triangle has an area of 1 square unit because it is half of a 2 by 1 rectangle. Similarly, the lower shaded triangle has an area of 1 1/2 square units because it is half of a 3 by 1 rectangle. Find the areas of these figures by subdividing them into squares, halves of squares, and triangles which are halves of rectangles.

   ★ a.   b.   ★ c.   d.

3. Sometimes it is easier to find the area outside a figure than inside. The hexagon in Figure **a** has been enclosed inside a square. What is the area of the shaded region? Subtract this area from the area of the 3 by 3 square to find the area of the hexagon. Use this technique to find the areas in Figures **b** through **d**.

   ★ a.   b.   ★ c.   d.

4. Find the area of each right triangle by enclosing it in the smallest possible rectangle. Record the areas of these rectangles and triangles in the accompanying table.

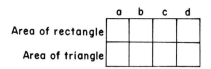

a.   b.   c.   ★ d.

e. Using *A* for area, *b* for base, and *h* for height or altitude, write a formula for the areas of the preceding triangles.

5. Compute the areas of the next 6 triangles by enclosing each in the smallest possible rectangle. Then label a base and altitude for each of these triangles and use the formula in Activity 4 to find their areas.

★ a.   b.   ★ c.

d.   ★ e.   f.

★ 6. There are 7 noncongruent triangles having an area of 1 square unit which can be formed on the geoboard. Sketch them here. (*Hint:* Use the formula for the area of a triangle.)

7. A parallelogram is a quadrilateral with two pairs of opposite parallel sides. A rectangle is a special case of a parallelogram. Find the areas of the three parallelograms by enclosing them in rectangles.

a.   b.   c.

★ These parallelograms have the same base, height (or altitude), and area as the rectangle above. What conjecture can be made from these examples? Test your conjecture on some other parallelograms.

★ 8. Find the areas of these trapezoids and write them in the table. Then compare the upper base (U), lower base (L), and height (H) of each trapezoid with its area to find a formula for the area of a trapezoid.

	H	U	L	Area
a	2	1	4	
b	2	1	3	
c	4	1	4	

a.   b.   c.

9. **Pick's Formula***: Any two polygons having the same number of boundary nails and the same number of interior nails will have the same area. Each polygon on these geoboards has 10 boundary nails and 4 interior nails.

★ a. Compute their areas.
   b. For each of the following tables form polygons having the given numbers of interior and boundary nails. Look for a relationship between the numbers in each table.

*For a development of Pick's formula, see G. Aman, "Discovery on a geoboard," *The Arithmetic Teacher,* **21** No. 4 (April 1974), pp. 267–72.

No interior nails	
Boundary nails	Area
3	
4	
5	
6	

One interior nail	
Boundary nails	Area
3	
4	
5	
6	

Two interior nails	
Boundary nails	Area
3	
4	
5	
6	

★ c. Use these tables to discover a formula for the areas of your polygons, in terms of the number of boundary nails (b) and the number of interior nails (i).

★ 10. **Mystery Polygon:** Form a polygon on your geoboard which satisfies each of the clues in **a** through **f**.

a. Nine nails on the boundary
b. Seven sides
c. Three nails in the interior
d. Two right angles
e. Area of 6 1/2 square units
f. Two sides of length 2 units

Just for Fun—Pentominoes

Pentominoes are polygons which can be formed by joining five squares along their edges. There are 12 such polygons. Trace these polygons and cut them out for the puzzles and the game.

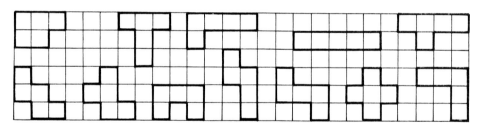

Puzzles: The 12 pentominoes can be pieced together into rectangular puzzles. The four which can be formed have these dimensions: 6 by 10, 5 by 12, 4 by 15, and 3 by 20.
★ Why are these the only possible rectangles? There are also several 8 by 8 square puzzles with four missing unit squares. In one of these the unit squares are missing from the corners, as shown here, and in another they are miss-
★ ing from the center of the puzzle. Try forming some of these puzzles.

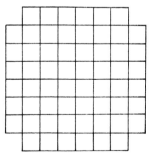

Pentomino Game: This game is played on an 8 by 8 checkerboard which should be constructed so that its squares are the same size as the unit squares of your pentominoes. The players take turns, each placing a pentomino on the uncovered squares of the board. Play continues until someone is unable to play (or all the pieces have been used). The player who has made the last move wins the game. The first seven plays of a game are shown on this board. Player A has made 4 moves, and player B has made 3 moves. Find a way player B can make the eighth move (with one of the remaining pentominoes) so that no more moves can be made. Play a few games and look for strategies.*

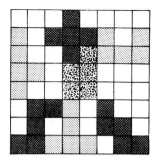

Polyominoes: Any number of unit squares can be placed edge to edge to form polygons. Two squares are called a domino. Polygons of three, four, and six squares are called trominoes, tetrominoes, and hexominoes. There are 35 hexominoes. Form these sets of figures and try making up puzzles, similar to the ones for pentominoes. In general, polygons formed by placing unit squares edge to edge are called *polyominoes*.

ACTIVITY SET 5.3

MODELS FOR VOLUME AND SURFACE AREA

"There are things which seem incredible to most men who have not studied mathematics."

Archimedes

One of the greatest mathematicians of all time was Archimedes (287–212 B.C.), a native of Syracuse on the island of Sicily. You may have heard the story of how Archimedes was able to solve King Hiero's problem, of whether or not his crown had been made of pure gold or a mixture of gold and silver. The king had weighed out the exact amount of gold for a crown, but upon receiving the crown was suspicious. The crown weighed as much as the original amount of gold, so there seemed to be no way of exposing the fraud. Archimedes was informed of the problem, and a solution occurred to him while he was taking a bath. He noticed that the amount of water which overflowed the tub depended on the extent his body was immersed. In the excitement to try his solution with gold and silver, Archimedes leaped from the tub and ran naked to his home, shouting, "Eureka! Eureka!"**

---

*For more Pentomino puzzles and games, see M. Gardner, "Mathematical Games," *Scientific American,* **198** No. 6 (December 1957), pp. 126–34; and **212** No. 4 (October 1965), pp. 96–104.

**J.R. Newman, *The World of Mathematics* (New York: Simon and Schuster, 1956), pp. 185–86.

As Archimedes discovered, the volume of a submerged object is equal to the volume of the displaced water. Finding the volumes of objects by submerging them in water is especially convenient with metric units, because the number of milliliters of water that is displaced is equal to the volume of the object in cubic centimeters. This relationship will be used in the following activities. You will also need the formulas for the volumes of prisms and cylinders (area of base × height), pyramids and cones (1/3 × area of base × height), and spheres (4/3 × πr^3).

★ 1. To understand Archimedes' solution consider two cubes both weighing the same amount, one of pure gold and one of pure silver. Since gold is about twice as heavy as silver, the cube of gold will be smaller than the cube of silver. Therefore, when each is submerged in water there will be more water displaced for the cube of silver.

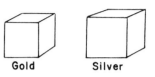
Gold   Silver

To apply his theory, Archimedes used the crown and an amount of pure gold weighing the same as the crown. He submerged them separately in water and measured the amount of overflow. There was more water displaced for the crown than for the pure gold. What does this show about the volume of the crown as compared to the volume of the gold? Was King Hiero cheated?

★ 2. Patterns for prisms, pyramids, and cylinders are easy to make. Three such patterns are contained on Material Card 26. Cut them out and tape their edges. Compute the volumes and surface areas of these figures. You may wish to fill them with sand, salt, etc., to check your answers. (The area of the base of a cylinder is $\pi \times r^2$. Use 3.14 for π.)

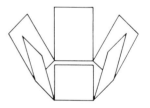

	Hexagonal Prism	Hexagonal Pyramid	Cylinder
Area of base			
Height (altitude)			
Volume			
Surface area			

★ 3. Make a cylinder (without bases) from a standard sheet of paper so that its circumference is 21.5 centimeters and its height is 28 centimeters. (There is a centimeter ruler on Material Card 41.) Then use another sheet of paper to form a cylinder with a circumference of 28 centimeters and a height of 21.5 centimeters. Compute their volumes. Are they equal?

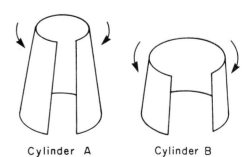
Cylinder A   Cylinder B

4. Cut a standard sheet of paper in half lengthwise and tape the two halves end to end, as shown next. Make a cylinder without bases and compute its volume. Compare the volume of this cylinder with that of Cylinder B in Activity 3. What happens to the volumes of cylinders as the height is halved and the diameter is doubled?

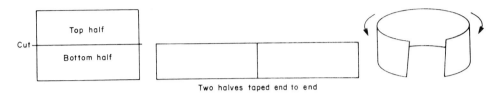

Two halves taped end to end

5. Make a set of six different cones by cutting out three discs, each of radius 10 cm. (There is a compass on Material Card 41.) Each disc can be used for making two cones. Cut sectors from these discs which have central angles of 45°, 90°, 135°, 225°, 270°, and 315°. (There is a protractor on Material Card 41.) After cutting each disc into two parts, bend and tape the edges to form the cones.

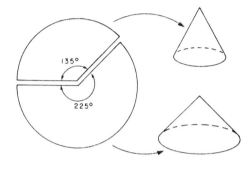

   a. Compare your cones. Without computing, select the one which you think has the greatest volume.

★ b. Write the volumes and base areas (to the nearest 2 decimal places) of the cones in the following table. (The area of the base of a cone is $\pi \times r^2$. Use 3.14 for π.)

★ c. What will happen to the volume of the cone as: sectors with central angles of less than 45° are used? sectors with central angles of greater than 315° are used?

Central angle of disc	45°	90°	135°	225°	270°	315°
Base area (cm^2)						
Height (cm)	9.92	9.68	9.27	7.81	6.61	4.84
Radius (cm)	1.25	2.50	3.75	6.25	7.50	8.75
Volume (cm^3)						

★ 6. Use one of your cones from Activity 5 and place a mark on the surface which is halfway from the vertex to the base. If the cone is filled to this mark, the volume of the contents will be what fractional part of the whole cone: 1/3, 1/4, 1/5, 1/6, 1/7, or 1/8? Take a guess and then fill your "half cone" with water several times to check your answer. Compute the volume of the "half cone" and compare it to the volume of the whole cone.

7. A basketball displaces 7235 milliliters of water.
   ★ a. What is its volume in cubic centimeters?
   b. What is the diameter of this ball? (Sphere volume = $4/3 \times \pi \times r^3$, where r is the radius. Use 3.14 for π.)
   ★ c. Submerge a baseball in water and measure the displaced water in milliliters. According to this experiment, what is the volume of a baseball in cubic centimeters? A baseball has a diameter of approximately 7.3 centimeters. Compute its volume in cubic centimeters and compare this with the results of your experiment.
   d. Submerge an unopened cylindrical can in water and measure the overflow. Compute the volume of the can and compare the answer with the results of your experiment.

8. Spherical objects such as Christmas tree ornaments and basketballs are packed in cube-shaped compartments and boxes.
   ★ a. If a sphere just fits into a box, what fraction of the box is "wasted space": 1/6, 1/5, 1/4, 1/3, or 1/2? Compute your answer for a sphere of diameter 25 centimeters.
   b. Spheres are also packed in cylinders. A cylinder is a better fit than a box but there is still extra space. If three balls are packed in a cylindrical can whose diameter equals that of a ball and whose height is three times the diameter, what fraction of the space is unused: 1/6, 1/5, 1/4, 1/3, or 1/2?
   ★ c. Tennis ball cans hold three balls. Determine the amount of wasted space in these cans if the balls have a diameter of 6.3 centimeters and the can has a diameter of 7 centimeters and a height of 20 centimeters. Check the reasonableness of your answer by experimenting with a can of balls and water.

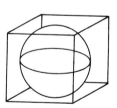

Just for Fun—Soma Cubes

Soma cubes is a seven-piece puzzle invented by the Danish author Piet Hein. Piece 1 has three cubes and the other six pieces each have four cubes. Surprisingly, these seven pieces can be assembled to form a 3 by 3 by 3 cube. You may wish to try this challenging puzzle. The soma cubes can be purchased or constructed by gluing together sugar cubes, wooden cubes, dice, etc.

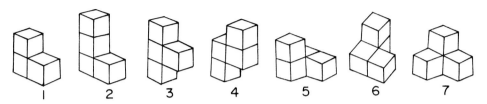

A common activity with soma cubes is constructing certain well-known figures. Here are three figures each of which requires all seven pieces.

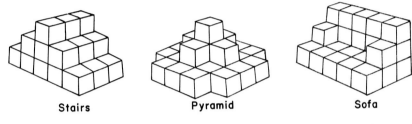

Stairs   Pyramid   Sofa

★ There are several elementary techniques for determining when a figure cannot be constructed with soma cubes. The simplest of these is counting the number of cubes. For example, a figure with 18 cubes cannot be formed because there is no combination of 4s and a 3 which adds up to 18. Use this approach to find which one of these two figures cannot be formed with soma cubes.*

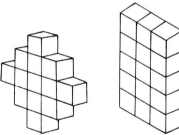

*Additional techniques can be found in G. Carson, "Soma Cubes," *The Mathematics Teacher,* **66** No. 7 (November 1973), pp. 583-92.

6

Fractions and Integers

ACTIVITY SET 6.1

MODELS FOR EQUALITY AND INEQUALITY

"Piaget has charted the cognitive development of pre-adolescents, and his research indicates that even at the age of twelve, most children deal well only with symbols that are closely tied to their perceptions. For example, the symbolic representation, 1/2 + 1/3 = 5/6, has meaning for most elementary school children only if they can relate it directly to concrete or pictorial representations."*

The second mathematics assessment (1978) of the National Assessment of Educational Progress (NAEP) found that 30 percent of the 13-year-olds added the numerators and denominators to find the sum of 1/2 and 1/3. A bit of reflection indicates that 2/5 is less than 1/2, one of the original addends.

"It seems apparent that more effort needs to be expended helping students internalize the concept of fraction. This can be done through careful, meaningful instruction. For through meaning, applications and transfer become realistic expectations rather than hoped-for outcomes."**

In the following activities you will use fraction bars and Cuisenaire rods to illustrate fractions. The fraction bars are on Material Cards 27, 28, and 29. The bars are divided into 2, 3, 4, 6, and 12 parts. The denominator is represented by the number of parts in a bar and the numerator by the number of shaded parts. The Cuisenaire rods are on Material Card 37. There are 10 different lengths of rods. Fractions are represented by comparing the lengths of two rods. The rods and bars should be cut out for use with the following activities.

1. The fractions for these bars are equal because both bars have the same amount of shading. In the set of 32 fraction bars (Material Cards 27, 28, and 29) there are three bars whose fractions equal 2/3. Sort your set of bars into piles so that the bars with the same shaded amount are in the same pile.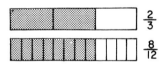

   ★ a. How many of these bars represent fractions which are equal to: 1/2, 2/3, 0/4, 6/6, 1/4?

   ★ b. Use the set of fraction bars to complete the following equalities.

   $$\frac{4}{6} = \frac{}{} = \frac{}{} \qquad \frac{9}{12} = \frac{}{} \qquad \frac{3}{6} = \frac{}{} = \frac{}{} = \frac{}{} \qquad \frac{1}{4} = \frac{}{}$$

---

*M.J. Driscoll, "The Role of Manipulatives in Elementary School Mathematics," *Research Within Reach: Elementary School Mathematics,* (St. Louis, Missouri: Cemrel, Inc., 1983), p. 1.

**T.R. Post, "Fractions: Results and Implications from National Assessment," *The Arithmetic Teacher,* No. 9 (May 1981), pp. 26–31.

c. Complete this table by writing in the sixths, fourths, thirds, and halves that are equal to the twelfths. (*Note:* Many of the squares should be left blank.)

Twelfths	$\frac{0}{12}$	$\frac{1}{12}$	$\frac{2}{12}$	$\frac{3}{12}$	$\frac{4}{12}$	$\frac{5}{12}$	$\frac{6}{12}$	$\frac{7}{12}$	$\frac{8}{12}$	$\frac{9}{12}$	$\frac{10}{12}$	$\frac{11}{12}$	$\frac{12}{12}$
Sixths													
Fourths													
Thirds													
Halves													

★ d. A fraction is not in *lowest terms* if the numerator and denominator have a common factor greater than 1. Circle all the fractions which are not in lowest terms. Each of these fractions which is greater than zero and less than 1 is equal to another fraction in the table which is in lowest terms.

2. By placing the five different types of fraction bars in a column, many equalities and inequalities can be seen. For example, by comparing the vertical lines of these bars we can see that:

$$\frac{1}{2} = \frac{6}{12} \qquad \frac{2}{3} < \frac{3}{4} \qquad \frac{5}{6} = \frac{10}{12}$$

★ a. List 10 different equalities of pairs of fractions which can be illustrated by comparing the vertical lines of these bars.

b. List 10 different inequalities of pairs of fractions which can be illustrated by these bars.

★ 3. Use your fraction bars for the following fractions: 5/6, 1/4, 11/12, 2/3, 1/6, 5/12, 3/4, 1/3, 7/12, 1/2, and 1/12. Place these bars in increasing order from the smallest shaded amount to the largest shaded amount and complete the following inequalities.

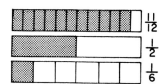

$$\rule{1em}{0.4pt} < \frac{1}{6} < \rule{1em}{0.4pt} < \rule{1em}{0.4pt} < \rule{1em}{0.4pt} < \frac{1}{2} < \rule{1em}{0.4pt} < \rule{1em}{0.4pt} < \rule{1em}{0.4pt} < \rule{1em}{0.4pt} < \frac{11}{12}$$

Each of the preceding fractions has a common denominator of 12. Rewrite these inequalities using only fractions with a denominator of 12.

$$\rule{1em}{0.4pt} < \frac{2}{12} < \rule{1em}{0.4pt} < \rule{1em}{0.4pt} < \rule{1em}{0.4pt} < \rule{1em}{0.4pt} < \rule{1em}{0.4pt} < \rule{1em}{0.4pt} < \rule{1em}{0.4pt} < \frac{11}{12}$$

★ 4. Each part of this 3/4 bar has been split into two equal parts. There are now eight parts and six of these are shaded. Since this splitting has neither increased nor decreased the total shaded amount of the bar, both 3/4 and 6/8 are fractions for the same amount.*

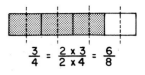

$$\frac{3}{4} = \frac{2 \times 3}{2 \times 4} = \frac{6}{8}$$

★   a. Split each part of these bars into two equal parts and complete the equations.

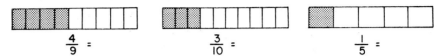

$$\frac{4}{9} = \qquad \frac{3}{10} = \qquad \frac{1}{5} =$$

★   b. Split each part of these bars into three equal parts and complete the equations.

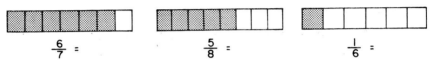

$$\frac{6}{7} = \qquad \frac{5}{8} = \qquad \frac{1}{6} =$$

★   c. Split each part of these bars into four equal parts and complete the equations.

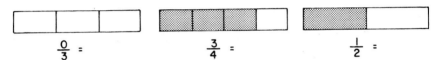

$$\frac{0}{3} = \qquad \frac{3}{4} = \qquad \frac{1}{2} =$$

★   d. Suppose that each part of the fraction bar for a/b is split into three equal parts. How many of the parts will be shaded? How many parts will there be in all? Write the fraction for this bar.

5. There are an infinite number of fractions that are equal to 2/3. These fractions can be generated by splitting the parts of a 2/3 bar.

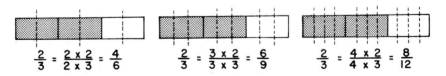

$$\frac{2}{3} = \frac{2 \times 2}{2 \times 3} = \frac{4}{6} \qquad \frac{2}{3} = \frac{3 \times 2}{3 \times 3} = \frac{6}{9} \qquad \frac{2}{3} = \frac{4 \times 2}{4 \times 3} = \frac{8}{12}$$

  a. If each part of a 2/3 bar is split into five equal parts, what fraction does the bar represent?

★   b. If each part of a 2/3 bar is split into n equal parts, what fraction does the bar represent?

★   c. What effect does splitting each part of a fraction bar into n equal parts have on the corresponding fraction (numeral) for the bar?

---

*For a similar approach to equivalent fractions, see H. Bohan, "Paper folding and equivalent fractions—bridging a gap," *The Arithmetic Teacher,* **18** No. 4 (April 1971), pp. 245-49.

d. Split each of the following bars to illustrate the given equalities.

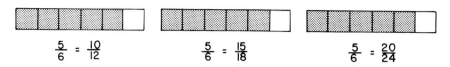

$$\frac{5}{6} = \frac{10}{12} \qquad \frac{5}{6} = \frac{15}{18} \qquad \frac{5}{6} = \frac{20}{24}$$

6. The parts of a 2/3 bar are bigger than the parts of a 3/4 bar. If each part of the 2/3 bar is split into 4 equal parts and each part of the 3/4 bar is split into 3 equal parts, both bars will then have 12 parts of equal size. The new fractions for these bars will have a common denominator of 12.

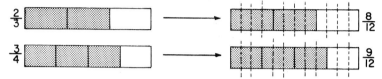

For each of the following pairs of bars, split the parts so that all the parts have the same size. The corresponding fractions will have the same denominator. Complete the equations.

★ a.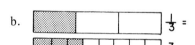

$\frac{1}{2} =$

$\frac{5}{7} =$

b.
$\frac{1}{3} =$

$\frac{3}{8} =$

c.
$\frac{4}{10} =$

$\frac{4}{6} =$

7. **Selecting a Unit:** The fraction represented by a Cuisenaire rod depends on the choice of the unit rod. If the dark green rod is the unit then the white rod represents 1/6, the red rod 1/3, the green rod 1/2, etc., as shown here.

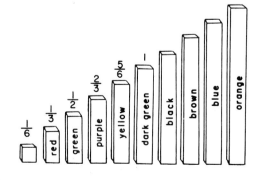

★ a. Each rod which is longer than the dark green rod represents a fraction whose numerator is greater than the denominator. Such fractions are sometimes called *improper fractions*.
Write the fractions for these rods in lowest terms.

108 CHAP. 6 FRACTIONS AND INTEGERS

b. A number which is expressed by a whole number and a fraction is called a *mixed number*. Write the mixed numbers for the four fractions in part a.

8. **Changing the Unit:** If the brown rod is the unit then the white rod represents 1/8.

   a. Write the fractions for the remaining 8 rods in lowest terms.

   ★ b. Write the lengths of the 2 longest rods as mixed numbers.

   c. If the orange rod is chosen as the unit, what is the fraction for the white rod?

   ★ d. If the green rod is 1/3, what is the unit rod?

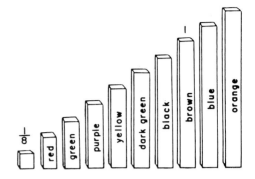

9. **Finding the Unit:** If the red rod is 1/5 then what rod represents 1/2? One way to answer this question is to place 5 red rods end to end to determine the unit. Then find the rod which is half as long. Another method is to find the rod which is equal to the length of 2 1/2 red rods.

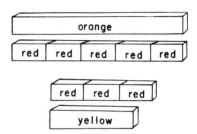

Use your rods to answer these questions. (*Hint:* The unit rod may be longer than an orange rod.)

   ★ a. If the dark green rod represents 3/4, what rod is 9/8?

   b. If the green rod is 1/4, what rod is 1/3?

   ★ c. If the yellow rod is 1/3, what rod is 1/5?

10. **Equal Fractions:** When fractions are represented by rods, equalities can be found by trains of the same length. If the unit rod is dark green then the all-white, all-red, and all-green trains show several pairs of equal fractions. As examples: 3/6 = 1/2, 4/6 = 2/3, and 2/6 = 1/3.

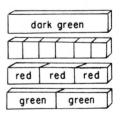

For each of the following rods, form all possible single-color trains having the same length. Write an equality of two fractions which can be determined by two of the trains.

   ★ a. 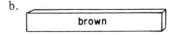   b.

ACTIVITY SET 6.1   109

★ c. [blue rod]   d. [orange rod]

11. **Different Rods for Fractions:** Depending on the unit chosen, several rods can be used to represent 1/3. One rod for 1/3 and its corresponding unit are shown here.

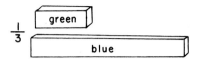

★ a. Name another 1/3 rod and the unit.

b. Name 3 rods representing 3/4, and their units. (The unit may be a train of 2 rods.)

★ c. Name 2 rods for representing 2/5, and their unit rods.

Just for Fun—Fraction Games (Fraction Bars—Material Cards 27, 28, and 29)

"FRIO"* (Inequality—2 to 4 players): Each player is dealt 5 bars in a row face up. These bars should be left in the order they are dealt. The remaining bars are placed in a stack face down.

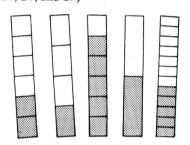

On a player's turn he/she takes a bar from the stack and uses it to replace any one of the 5 bars. The object of the game is to get 5 bars in order from the *smallest* shaded amount to the *largest* or from the *largest* to the *smallest*. In this example there will be 5 bars in order if the 2/6 bar is replaced by a whole bar and the 1/4 bar is replaced by an 11/12 bar.

The first player to get 5 fraction bars in decreasing or increasing order wins the game.

"Fraction Bingo" (Equality—2 to 4 players): Each player selects a card from those shown next. The deck of 32 bars is spread face down. On a player's turn he/she takes a bar and circles the fraction or fractions on his/her mat which equal(s) the fraction from the bar. The first player to circle four fractions in any row, column, or diagonal is the winner. (Strategy: If the bars are colored green, yellow, blue, red, and orange for halves, thirds, fourths, sixths, and twelfths, respectively, you can increase your chances of winning by selecting bars of the appropriate color.)

Card 1				Card 2				Card 3				Card 4			
$\frac{4}{6}$	$\frac{1}{6}$	$\frac{0}{4}$	$\frac{11}{12}$	$\frac{1}{3}$	$\frac{1}{4}$	$\frac{1}{6}$	$\frac{0}{3}$	$\frac{2}{3}$	$\frac{1}{4}$	$\frac{2}{6}$	$\frac{6}{12}$	$\frac{1}{4}$	$\frac{1}{6}$	$\frac{7}{12}$	$\frac{1}{2}$
$\frac{5}{12}$	$\frac{1}{4}$	$\frac{5}{6}$	$\frac{3}{6}$	$\frac{7}{12}$	$\frac{0}{6}$	$\frac{1}{12}$	$\frac{4}{6}$	$\frac{5}{6}$	$\frac{0}{3}$	$\frac{5}{12}$	$\frac{1}{6}$	$\frac{0}{3}$	$\frac{5}{12}$	$\frac{2}{3}$	$\frac{4}{4}$
$\frac{3}{4}$	$\frac{1}{12}$	$\frac{1}{3}$	$\frac{11}{12}$	$\frac{2}{3}$	$\frac{3}{4}$	$\frac{5}{6}$	$\frac{5}{12}$	$\frac{7}{12}$	$\frac{4}{4}$	$\frac{1}{2}$	$\frac{11}{12}$	$\frac{1}{12}$	$\frac{5}{6}$	$\frac{11}{12}$	$\frac{2}{12}$
$\frac{6}{12}$	$\frac{4}{4}$	$\frac{9}{12}$	$\frac{10}{12}$	$\frac{11}{12}$	$\frac{2}{2}$	$\frac{3}{6}$	$\frac{1}{2}$	$\frac{3}{4}$	$\frac{8}{12}$	$\frac{1}{4}$	$\frac{1}{3}$	$\frac{6}{6}$	$\frac{3}{4}$	$\frac{0}{12}$	$\frac{1}{3}$

*R. Drizigacker, "FRIO, or FRactions In Order." *The Arithmetic Teacher,* **13** No. 8 (December 1966), pp. 684-85.

ACTIVITY SET 6.2

COMPUTING WITH FRACTION BARS

"Instead of presenting mathematics as rigorously as possible, present it as intuitively as possible. In place of abstractions, we must, as far as possible present concrete materials."
                                                                                    Morris Kline

National and statewide assessment results and related research indicate that many students of junior high school age have little conceptual understanding of fractions. They seem to operate on symbols without an adequate quantitative basis for their thinking. This is illustrated by the responses of 13-year-olds to the following test question from the second mathematics assessment of the National Assessment of Educational Progress (NAEP). Explain how answers of 19 and 21 could have been obtained.

Estimate the answer to 12/13 + 7/8. You will not have time to solve the problem using paper and pencil.

Responses	Percent Responding, Age 13
○ 1	7
● 2	24
○ 19	28
○ 21	27
○ I don't know	14

"Students appear to be learning many mathematical skills at a rote manipulation level and do not understand the concepts underlying the computation. . . . It appears that if children have not mastered a computational algorithm, they cannot use intuitive models to solve simple problems. The extremely poor performance in estimating the answer to exercises like 12/13 + 7/8 indicates that too frequently computational algorithms have little meaning, even for students who can successfully apply them."*

In this activity set you will find that by using the basic meanings of addition, subtraction, multiplication, and division, it is possible to compute with fractions in a visual and intuitive manner without using the standard rules for these operations. This approach will help to provide insight and meaning to these operations. The set of bars on Material Cards 27, 28, and 29 will be needed for some of these activities.

---

*T.P. Carpenter, H. Kepner, M.K. Corbitt, M.M. Lindquist, and R.E. Reys, "Results and Implications of the Second NAEP Mathematics Assessment: Elementary School," *The Arithmetic Teacher*, No. 8 (April 1980), pp. 10–12, 44–47.

1. **Addition and Subtraction**: Write the missing fractions and the sum or difference of the fractions for each pair of bars. Determine your answers by visual inspection of the bars and then check your results by computation. For sums greater than 1, write your answer as a mixed number.

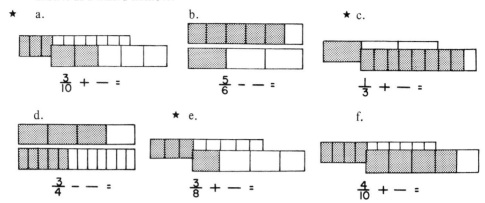

★ a. $\frac{3}{10} + — =$

b. $\frac{5}{6} - — =$

★ c. $\frac{1}{3} + — =$

d. $\frac{3}{4} - — =$

★ e. $\frac{3}{8} + — =$

f. $\frac{4}{10} + — =$

2. Turn your set of bars face down and select any five pairs. Complete the following equations for the sum and difference of each pair of fractions. (Subtract the smaller from the larger if the fractions are unequal.)

3. **Addition and Subtraction (Obtaining common denominators)**: If each part of the 2/3 bar is split into four equal parts and each part of the 1/4 bar is split into three equal parts, both bars will have 12 parts of the same size. These new bars show that $2/3 + 1/4 = 11/12$ and $2/3 - 1/4 = 5/12$.

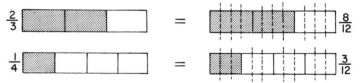

Split the parts of the following fraction bars so that each pair of bars has the same number of parts of the same size.

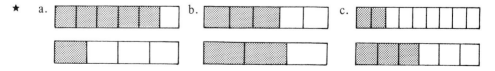

★ a. b. c.

112 CHAP. 6 FRACTIONS AND INTEGERS

Use these bars to compute the following sums and differences.

★ d. $\dfrac{5}{6} + \dfrac{1}{4} =$ e. $\dfrac{3}{5} + \dfrac{2}{3} =$ f. $\dfrac{2}{9} + \dfrac{3}{6} =$

★ g. $\dfrac{5}{6} - \dfrac{1}{4} =$ h. $\dfrac{2}{3} - \dfrac{3}{5} =$ i. $\dfrac{3}{6} - \dfrac{2}{9} =$

★ 4. **Multiplication:** One-third times 1/6 can be determined by splitting the shaded part of a 1/6 bar into three equal parts. One of these smaller parts is 1/18 of a bar because there are 18 of these parts in a whole bar.

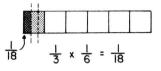

★ a. Split the shaded amount of each bar into two equal parts. Use this result to complete the given equations.

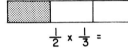  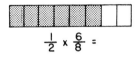

★ b. Split the shaded amount of each bar into three equal parts and complete the equations.

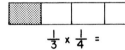 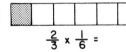 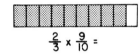

★ c. Split the shaded amount of each bar into four equal parts and complete the equations.

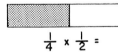 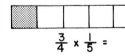 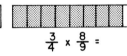

5. **Division:** The shaded amount of the 2/3 bar is four times the shaded amount of the 1/6 bar. Viewed in another way, we can say that one shaded amount "exactly fits" into the other four times.

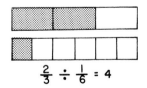

Determine the following quotients by comparing each pair of bars. Then check your results by using the formula for division: $a/b \div c/d = a/b \times d/c$.

a. 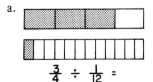   b. 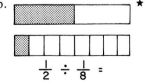   ★ c.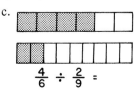

ACTIVITY SET 6.2   113

6. Spread the bars face down and shuffle them. Turn over two bars and compare their shaded amounts. If the quotient of the two fractions is a whole number, record the result by writing a division equation. If the quotient is not a whole number, continue selecting bars, one at a time, until a whole number quotient can be formed. Remove the two bars which have been used. If two zero bars are selected, you cannot write an equation because the quotient is not defined. Continue this activity until there are no bars left. See how many whole number quotients you can get.

─ ÷ ─ =   ─ ÷ ─ =   ─ ÷ ─ =   ─ ÷ ─ =

─ ÷ ─ =   ─ ÷ ─ =   ─ ÷ ─ =   ─ ÷ ─ =

The preceding activity can be played like a solitaire game, by playing through the deck of bars several times and keeping a record of your scores.

★ a. What is the greatest possible number of whole number quotients from the 32 bars?

b. If this game were scored by adding up the whole number quotients, what is the greatest possible score? (*Hint:* Match the largest fractions with the smallest.)

c. What is a reasonable score to try to beat, for winning the game suggested in part **b**? This can be determined by playing a few games.

7. **All Four Operations:**

★ a. The following ten fractions are from the fraction bars.

$$\frac{1}{6} \quad \frac{2}{4} \quad \frac{5}{12} \quad \frac{2}{6} \quad \frac{0}{3} \quad \frac{3}{4} \quad \frac{3}{12} \quad \frac{2}{3} \quad \frac{1}{2} \quad \frac{1}{4}$$

Try placing these fractions in the 10 blanks to form four equations.

─ + ─ = ─   3 × ─ = ─

─ − ─ = ─   ─ ÷ ─ = 2

b. Spread your bars face down and select any ten of them. Using the fraction from each bar only once, complete as many of these four equations as possible. If you cannot form all four equations, continue to select bars one at a time until all the equations can be completed.

─ + ─ = ─   3 × ─ = ─

─ − ─ = ─   ─ ÷ ─ = 2

c. **Solitaire:** The activity in part **b** can be played like a solitaire game by seeing how many turns it takes you to complete the four equations by selecting only 10 bars on each turn.

Just for Fun—Fraction Games for Operations (Fraction Bars—Material Cards 27, 28, and 29)

Fraction Bar Black Jack (Addition—2 to 4 players): Spread the bars face down. The object is to select one or more bars so that the fraction or the sum of fractions is as close to 1 as possible, but not greater than 1. Each player selects his/her own bars by taking one at a time. (You may wish to take only 1 bar.) A player finishes his/her turn by saying, "I'm holding." After every player has finished, the players show their bars. The player who is closest to a sum of 1, but not over, wins the round.

Examples Player 1 has a sum greater than 1 and is "over." Player 2 has a greater sum than the fraction from Player 3's bar, and wins the round.

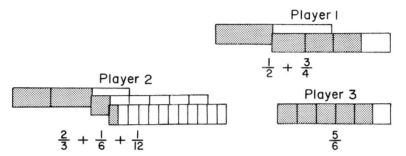

Solitaire (Subtraction): Spread the fraction bars face down. Turn over 2 bars and compare their shaded amounts. If the difference between the two fractions is less than 1/2, you win the 2 bars. If not, you lose the bars. See how many bars you can win by playing through the deck. For the two pairs of bars shown here, will you win the top two or the bottom two?

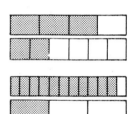

Greatest Quotient (Division—2 to 4 players): Remove the zero-bars from your set of 32 bars and spread the remaining bars face down. Each player takes 2 bars. The object of the game is to get the greatest possible quotient by dividing one of the fractions by the other. The greatest whole number of times that one fraction divides into the other is the player's score. Each player has the option of taking another bar to improve his/her score or passing. If he/she takes another bar, 1 bar must be discarded from the hand. The first player to score 21 points wins the game.

Examples The whole number part of the quotient can be determined by comparing the shaded amounts of the bars. The score for the top 2 bars is 4, because the shaded amount of the 2/3 bar is four times greater than the shaded amount of the 1/6 bar. The score for the bottom 2 bars is 1, because the quotient is greater than 1 but less than 2.

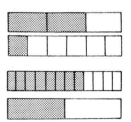

ACTIVITY SET 6.3

MODELS FOR OPERATIONS WITH INTEGERS

The rules for computing with positive and negative numbers, also referred to as *signed numbers*, developed over hundreds of years from trading and commerce. We will use similar trading activities with black and red chips to illustrate addition, subtraction, multiplication, and division of integers. Black chips will represent positive numbers, and red chips will represent negative numbers. By agreeing that each black chip cancels a red chip, every integer can be represented in many ways. For example, a set of 3 black chips and 5 red chips represents ⁻2. If there are as many black chips as red chips in a set, the set represents zero. In this case the positive number from the black chips and the negative number from the red chips are called *opposites* (also *negatives* or *inverses*) of each other. Cut out the chips on Material Card 30 for the activities on the following pages.

1. Two thousand years ago the Chinese were using black and red rods to represent signed numbers. The trading activity which is described by the following list of transactions is similar to one which might have occurred in that country. Imagine that you are doing the trading and that you receive *black chips* for objects you *sell* and *red chips* for objects you *buy*. Let each chip correspond to a yen. After you have gone through this list, match your black chips with your red chips to determine how many yen you should pay or receive.

    List of Transactions
    Sell 1 dozen rice cakes for 3 yen; sell 6 bundles of wood for 2 yen; buy 1 blanket for 4 yen; sell 8 straw hats for 2 yen; buy 3 fans for 1 yen; buy 1 rickshaw bumper sticker for 1 yen; buy 1 Buddha statue for 2 yen; sell 1 bucket of goat's milk for 1 yen; buy 1 basket of rice seed for 1 yen; sell 1 box of cheese for 3 yen; buy 4 pairs of sandals for 4 yen.

    ★ a. What is the total number of black chips you received?

    b. What is the total number of red chips you received?

    ★ c. Should you pay or will you be paid? How much?

2. After a day of trading, Yung, Ky, and Ming each have the sets of chips shown next. Yung's chips represent the number ⁻3, since after pairing black and red chips there are 3 red chips left. What numbers are represented by the other sets of chips?

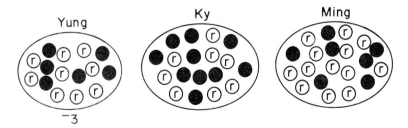

Addition:

★ 3. Yung collected two sets of chips, one from trading in the morning and the other from the afternoon. Under each of these sets write the number represented by the chips.

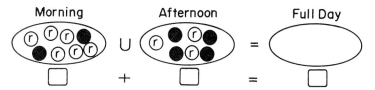

a. The union of these two sets is the set of chips for the full day. Sketch these chips in the above diagram. Write the number in the box represented by this set.

b. Sketch the smallest set of chips which represents the full day's total.

4. In parts **a**, **b**, and **c**, form two sets which have the indicated numbers of chips. Take their union and cancel pairs of black and red chips. Use the remaining chips to determine the given sums.

★   a. 7 black chips   b. 4 red chips   c. 8 red chips
       3 red chips   3 red chips   2 black chips

       $7 + {}^-3 =$ ${}^-4 + {}^-3 =$ ${}^-8 + 2 =$

★   d. State a rule for adding a positive and negative integer.

     e. State a rule for adding two negative integers.

Subtraction:*

5. By 10 A.M. Ming and Ky had identical sets of chips each representing $^-5$. Then Ming decided to return a box of tea which had been purchased earlier, and she had 2 red chips taken away from her collection. Remove 2 red chips and complete the equation below Ming's set. Shortly thereafter, Ky sold a chicken and received 2 black chips. Put 2 black chips into Ky's set and complete the equation.

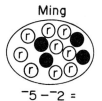 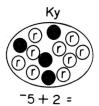

★   a. What are the new numbers represented by Ming's and Ky's chips?

     b. What generalization do these transactions suggest about taking out red chips as compared to putting in black chips?

---

*For further illustrations with the black and red chips model, see A.B. Bennett, Jr. and G.L. Musser, "A concrete approach to integer addition and subtraction," *The Arithmetic Teacher,* **23** No. 5 (May 1976), pp. 332-36.

★ c. State the generalization in part **b** in terms of subtracting negative numbers and adding positive numbers.

6. Sets A and B show how we can compute $5 - 8$. Since we can't take away 8 black chips from set A, set A is changed to set B by putting in 3 more black chips and 3 red chips. Set B still represents the number 5, but now 8 black chips can be removed, leaving 3 red chips. Therefore, $5 - 8 = {}^-3$.

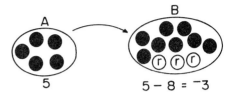

Use your chips to form the following sets. Then change each set to a new set by putting in more chips so that the appropriate number of chips can be taken away. Use these results to complete the equations. In each example sketch the new set.

★ a. Change this set so that 3 black chips can be taken away.

b. Change this set so that 2 red chips can be taken away.

c. Change this set so that 5 black chips can be taken away.

d. Explain how the activities in parts **a, b,** and **c** can be used to show that subtraction can be done by adding opposites. That is, $a - b = a + {}^-b$.

Multiplication:

7. The chips can be used to illustrate a model for multiplication if we agree that in the product $n \times s$, n tells the number of times we *put in* or *take out* s chips. For example, ${}^-2 \times 3$ means that 2 times we take out 3 black chips. This changes set A which represented 0 to set B which represents ${}^-6$. This shows that ${}^-2 \times 3 = {}^-6$.

Each set shown next represents zero. Make the given changes in these sets by sketching in or crossing out chips. Use the results to complete the equations.

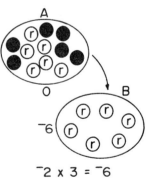

118   CHAP. 6   FRACTIONS AND INTEGERS

a. Three times put in 2 red chips. The number representing this set will then be ___.

★ b. Two times take out 4 red chips. The number representing this set will then be ___.

c. Three times take out 2 black chips. The number representing this set will then be ___.

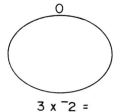

$3 \times {}^-2 =$

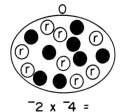

${}^-2 \times {}^-4 =$

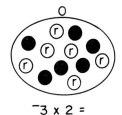

${}^-3 \times 2 =$

Division:

8. At the end of the day Ming had 6 red chips. Her three children said they would each pay an equal share of her debts. Divide this set into three equal parts. How many chips did each child get? Use this result to complete the equation.

${}^-6 \div 3 =$

9. Divide the following sets into the given numbers of parts and complete the equations. For example, the divisor in part **a** is 4 so the set of chips should be divided into four equal parts. These examples illustrate the *partitive use* of division.

★ a. Divide into 4 equal parts.

b. Divide into 3 equal parts.

c. Divide into 5 equal parts.

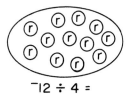

${}^-12 \div 4 =$

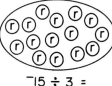

${}^-15 \div 3 =$

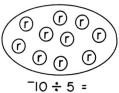

${}^-10 \div 5 =$

d. How many times can 3 red chips be taken away from this set of chips? Use your answer to complete the equation. This example illustrates the *subtractive* or *measurement use* of division.

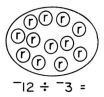

${}^-12 \div {}^-3 =$

Just for Fun—Games for Negative Numbers

The following four games require two dice, one white and one red. The white die represents positive numbers and the red die represents negative numbers.

Game 1 (Addition—2 to 4 players): Each player begins the game with a marker on the zero point of the number line. On a player's turn the dice are rolled and the positive

ACTIVITY SET 6.3

number from the white die is added to the negative number from the red die. This sum determines the amount of movement on the number line. If it is positive, move to the right, and if it is negative, move to the left. If the sum is zero, roll again. The first player to reach either ⁻10 or 10 wins the game. If this game is played by moving the marker on each turn for the two dice separately, to the right for the white die and to the left for the red die, it will quickly lead to the rule for adding positive and negative numbers.

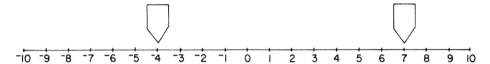

Game 2 (Addition and Subtraction—2 to 3 players): Each player selects one of the three cards shown next. On a player's turn the dice are rolled and the two numbers are added or subtracted. Suppose, for example, that you rolled 4 and ⁻6: you could add 4 + ⁻6 = ⁻2; or subtract, 4 − ⁻6 = 10, or ⁻6 − 4 = ⁻10. One of these numbers is then marked on the player's card. The first player to complete a row, column, or diagonal wins the game.

Card 1

⁻9	⁻11	⁻1	⁻8	10
12	3	⁻10	⁻2	⁻6
8	⁻12	⁻7	⁻4	⁻5
2	6	11	7	1
⁻3	0	5	9	4

Card 2

5	12	10	⁻1	⁻9
⁻10	⁻3	7	11	⁻8
⁻4	2	4	⁻6	⁻2
9	⁻7	⁻5	⁻11	1
8	6	3	0	⁻12

Card 3

⁻1	⁻7	9	4	⁻6
8	⁻5	11	⁻12	6
⁻11	1	⁻2	7	⁻9
⁻3	10	12	⁻4	3
2	⁻8	5	0	⁻10

Game 3 (Multiplication and Inequality—2 to 4 players): Each player on his/her turn rolls the dice and multiplies the positive number from the white die by the negative number from the red die. This is repeated a second time and the player selects the greater of the two products. In this example, each player has had four turns. The greater product in each case has been circled. Player B is ahead at this point with the greater score: a total of ⁻12. After ten rounds the player with the greater sum wins the game.

Player A
(⁻4)   ⁻6
(⁻2)   ⁻18
(⁻6)   ⁻30
 ⁻15   (⁻6)
Total ⁻18

Player B
(⁻1)   ⁻3
 ⁻6   (⁻2)
(⁻5)   ⁻16
 ⁻24   (⁻4)
Total ⁻12

Chance Option: If a player obtains two undesirable numbers, such as ⁻25 and ⁻15, he/she may wish to use the *chance option*. This is done by rolling the dice a third time. If this product is greater than the other two, say, ⁻12, the player may select this number. If the third product is less than or equal to either of the other two, say, ⁻18, then the smallest of the three numbers (⁻25 in this example) must be kept as the player's score.

Game 4 (All four basic operations—1 to 4 players): Roll the dice four times and record the four negative numbers from the red die and the four positive numbers from the white die. Write these eight numbers in the blanks of the equations under *Round 1* so that you complete as many equations as possible. Repeat this activity for *Round 2* and *Round 3*. You receive one point for each equation. If, however, you complete all four equations for a given round, you receive 4 points plus 2 bonus points. The total number of points for the three rounds is your score.

Round 1	Round 2	Round 3
___ + ⁻1 = ___	3 + ___ = ___	___ + ___ = ⁻6
___ − ⁻4 = ___	⁻1 − ___ = ___	___ − ⁻5 = ___
⁻2 × ___ = ___	⁻1 × ___ = ___	3 × ___ = ___
___ ÷ 3 = ___	___ ÷ ___ = 2	___ ÷ ⁻1 = ___

7

Decimals: Rational and Irrational Numbers

ACTIVITY SET 7.1

MODELS FOR DECIMALS

"One of the major reasons children learn patterns of error is that their teachers have introduced them to paper-and-pencil procedures while they still need to work on problems with concrete aids."*

Robert B. Ashlock

The results of the second mathematics assessment (1978) of the National Assessment of Educational Progress (NAEP) contain some important implications for teaching decimals. "In analyzing the computational errors made on the decimal exercises it is evident that much of the difficulty lies in a lack of conceptual understanding. . . . It is important that decimals be thought of as numbers and the ability to relate them to models should assist in understanding."**

Decimal squares will be used as the model for decimals in this activity set. These squares are partitioned into 10, 100, and 1000 equal parts to illustrate one-place, two-place, and three-place decimals.

1. **Part-to-Whole Concept:** This square has 1000 equal parts and 80 parts are shaded. The decimal for this square is .080. Shade each square below so that the decimal tells how much of the square is shaded.

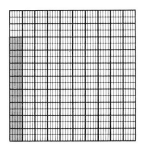

★   a. Hundredths Square   b. Tenths Square   c. Thousandths Square

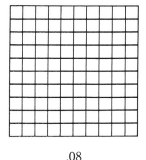 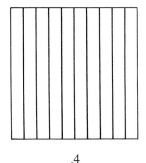 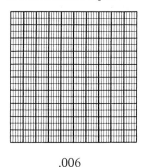

   .08   .4   .006

---

*R.B. Ashlock, *Error Patterns in Computation: A Semi-Programmed Approach.* (Columbus, Ohio: Charles E. Merrill, 1976).

**T.P. Carpenter, M.K. Corbitt, H.S. Kepner, M.M. Lindquist, and R.E. Reys, "Decimals: Results and Implications from National Assessment," *The Arithmetic Teacher,* No. 4 (April 1981), pp. 34–37.

★ d. Which of these decimals, .5, .05, or .50 represents 5 shaded parts out of 100? Describe the decimal square for the other two decimals.

e. Which of these decimals, .040, .40, or .004 represents 4 shaded parts out of 1000? Describe the decimal square for the other two decimals.

2. **Part-to-Whole Concept:** Write the decimal that represents the shaded amount for each of these squares.

★ a.   b.   c.

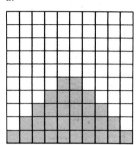   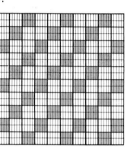   

_____   _____   _____

What decimal represents each of the following squares?
d. 9 shaded parts out of 10
★ e. 10 shaded parts out of 10
f. 9 shaded parts out of 100
g. 100 shaded parts out of 100
★ h. 90 shaded parts out of 1000

3. **Equality:** The decimals for these squares are equal because each square is the same size and has the same amount of shading. That is, 4 parts out of 10 is equal to 40 parts out of 100, and 40 parts out of 100 is equal to 400 parts out of 1000.

.4 = .40 = .400

124 CHAP. 7 DECIMALS: RATIONAL AND IRRATIONAL NUMBERS

Fill in the boxes below each square to complete the statement and the equation.

★ a.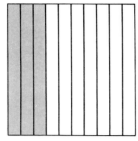

b.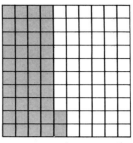

★ c.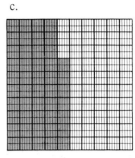

3 parts out of 10 is equal to ☐ parts out of 100.
.3 = ☐

42 parts out of 100 is equal to ☐ parts out of 1000.
.42 = ☐

470 parts out of 1000 is equal to ☐ parts out of 100.
.470 = ☐

d.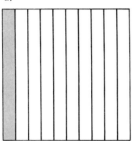

★ e.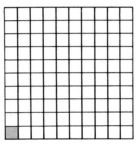

f.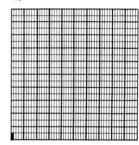

1 part out of 10 is equal to ☐ parts out of 100.
.1 = ☐

1 part out of 100 is equal to ☐ parts out of 1000.
.01 = ☐

1 part out of 1000 is equal to ☐ parts out of 10,000.
.001 = ☐

4. **Place Value:** The encircled parts of the square for .435 show that .435 can be thought of as 4 tenths, 3 hundredths, and 5 thousandths. Because of this, we say that 4 is in the *tenths place,* 3 is in the *hundredths place,* and 5 is in the *thousandths place.*

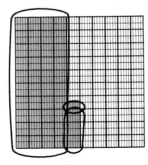

Tenths	Hundredths	Thousandths
4	3	5

ACTIVITY SET 7.1   125

Fill in the boxes below and encircle the parts of the square to illustrate each statement. Write the digits for each decimal in the place value table.

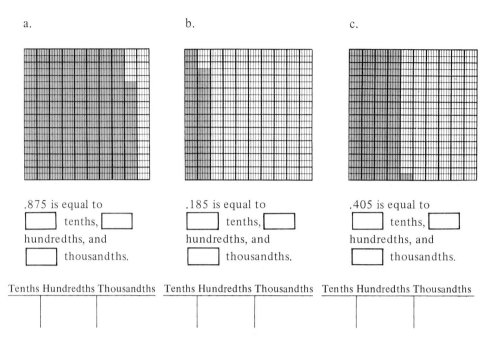

a.

.875 is equal to
☐ tenths, ☐ hundredths, and ☐ thousandths.

Tenths Hundredths Thousandths

b.

.185 is equal to
☐ tenths, ☐ hundredths, and ☐ thousandths.

Tenths Hundredths Thousandths

c.

.405 is equal to
☐ tenths, ☐ hundredths, and ☐ thousandths.

Tenths Hundredths Thousandths

5. **Regrouping:** In Figure A the parts of the square for .48 have been grouped into 4 tenths (40 hundredths) and 8 hundredths. In Figure B the parts of the square for .48 have been regrouped into 3 tenths (30 hundredths) and 18 hundredths.

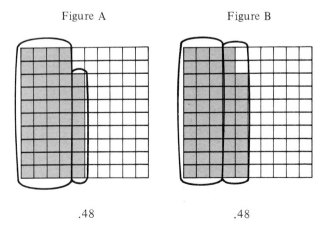

Figure A
.48

Figure B
.48

126 CHAP. 7 DECIMALS: RATIONAL AND IRRATIONAL NUMBERS

Fill in the boxes in the statements below and encircle the parts of each square to illustrate the regrouping.

★ a.

b.

c.

7 tenths and 2 hundredths regroup to 6 tenths and ☐ hundredths.

3 tenths, 6 hundredths, and 5 thousandths regroup to 2 tenths, ☐ hundredths, and 5 thousandths.

8 tenths, 6 hundredths, 5 thousandths regroup to 8 tenths, ☐ hundredths, and 15 thousandths.

d. The shaded parts of the decimal square for .48 are grouped in two different ways in Activity 5. How many different ways can .48 be grouped in terms of tenths and hundredths? List them.

Tenths	Hundredths
4	8
3	18

★ e. How many different ways can the shaded parts of the decimal square for .215 be grouped in terms of tenths, hundredths, and thousandths? List them.

6. **Regrouping:** In Figure C the parts of the square for .36 have been grouped into 2 tenths (20 hundredths) and 16 hundredths. In Figure D the parts of the square for .36 have been regrouped into 3 tenths (30 hundredths) and 6 hundredths.

Figure C

Figure D

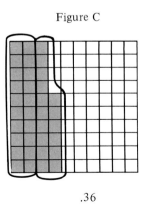

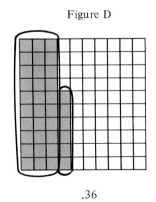

.36

.36

ACTIVITY SET 7.1

Fill in the boxes in the statements below and encircle the parts of each square to illustrate the regrouping.

★ a.

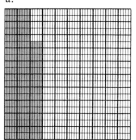

2 tenths, 6 hundredths, and 15 thousandths regroup to 2 tenths, ☐ hundredths, and 5 thousandths.

b.

8 tenths and 16 hundredths regroup to ☐ tenths and 6 hundredths.

c.

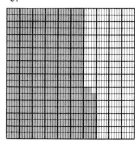

5 tenths, 13 hundredths, and 8 thousandths regroup to ☐ tenths, 3 hundredths, and 8 thousandths.

7. **Part-to-Whole Concept:** If each part of a decimal square for tenths is partitioned into 10 equal parts, there will be 100 parts. One of these small parts is .01 of the whole square.

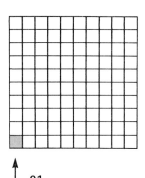

↑
└ .01

a. How many parts will there be if each part of a decimal square for hundredths is partitioned into 10 equal parts? What is the decimal for one of these small parts?

★ b. How many parts will there be if each part of a decimal square for thousandths is partitioned into 10 equal parts? What is the decimal for one of these small parts?

c. How many parts will there be if each part of a decimal square for thousandths is partitioned into 1000 equal parts? What is the decimal for one of these small parts?

8. **Inequality:** The numbers .6, .62, and .610 have one, two, and three decimal places, respectively. By partitioning the decimal square for .62 into 1000 equal parts we can see that .62 is equal to .620.

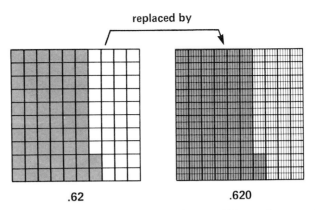

.62   replaced by   .620

★ a. If the decimal square for .6 is partitioned into 1000 equal parts, what decimal will represent the square? Which of the decimals .6, .62, or .610 is the greatest?

b. Write each of the following decimals as decimals with three places and then circle the greatest decimal and underline the smallest decimal.

                 .3   .27   .298

★ c. Write each of the following decimals as decimals with four places and then circle the greatest decimal and underline the smallest decimal.

                 .042   .0047   .04

d. Describe an easy method of ordering decimal numbers.

e. The following question is from a 1979 mathematics assessment test that was given by the City University of New York to entering freshmen. Less than 30% of 7100 students answered this question correctly.*

     Question: Which of the following numbers is the smallest?
     (a) .07   (b) 1.003   (c) .08   (d) .075   (e) .3

Use your method from part **d** to answer this question.

9. **Approximation:**

a. Using a decimal square for tenths, how many parts should be shaded to best approximate 1/3 of the square? Shade these parts. What decimal represents the shaded area?

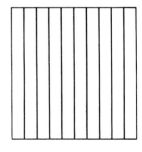

---

*A. Grossman, "Decimal Notation: An Important Research Finding," *The Arithmetic Teacher,* No. 9 (May 1983), pp. 32–33.

★ b. Shade the parts of each of these two decimal squares to best approximate 1/3 of the square. Write the decimal which represents the shaded area of each square.

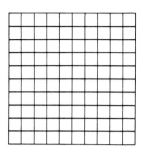 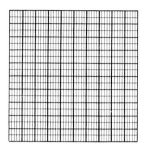

_____   _____

c. Shade the parts of each of these decimal squares to best approximate 1/6 of the square. Write the decimal which represents the shaded area of each square.

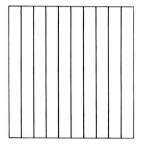 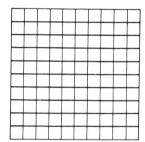 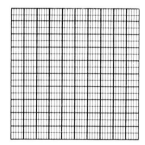

_____   _____   _____

Just For Fun—**Decimal Games** (Decimal Squares—Material Cards 32, 33, 34, and 35)

Decimal Bingo: (Equality, 2-4 players): Each player selects one of the four cards of decimals shown below. The deck of decimal squares should be spread face down. On a player's turn he/she takes a square and circles the decimal or decimals on her/his mat which equals the decimal for the square. The first player to circle four decimals in any row, column, or diagonal is the winner.

Card 1

.30	.400	.1	.70
.7	.55	.3	.8
.10	.40	.85	.95
.8	.300	.05	.550

Card 2

.60	.50	.3	.10
.1	.75	.600	.5
.450	.30	.15	.525
.500	.45	.425	.750

Card 3

.70	.2	.50	.40
.5	.4	.450	.65
.20	.700	.375	.675
.400	.45	.475	.650

Card 4

.75	.300	.60	.4
.600	.9	.30	.350
.40	.750	.325	.25
.35	.6	.575	.90

130   CHAP. 7   DECIMALS: RATIONAL AND IRRATIONAL NUMBERS

Decimal Place Value Game: (Place value, 2-4 players) Each player should draw a place value table like the one shown here. Each player in turn rolls two dice until obtaining a sum less than 10. The player then records this number in any one of the three places in her/his table. Only one number should be written in each place in the table, and the number cannot be changed to another column once it is written. After each player has had three turns, the player with the greatest three-place decimal is the winner. The first player to win five rounds wins the game.

Variation: Use a place value table with four columns and roll the dice to obtain the digits for a four-place decimal.

ACTIVITY SET 7.2

OPERATIONS WITH DECIMAL SQUARES

> "We should be spending more time having children become familiar with decimals, their meanings and uses, before rushing directly to decimal computation. Think of the time we spend with counting objects and modeling whole numbers, and writing these numbers before formal operations with whole numbers are introduced."*

"Symbols become appropriate after their associated concepts have been abstracted."** This is true for the symbols (numerals) for decimals as well as the symbols for decimal operations. In this activity set you will use the Decimal Squares model to illustrate the four basic operations with decimals. This model will enable you to understand why there is a close relationship between the operations with whole numbers and the corresponding operations with decimals.

1. **Addition:** Addition of decimals can be illustrated by the total shaded amount of two squares. Fill in the missing number in the box and compute the sum by counting the total number of shaded parts.

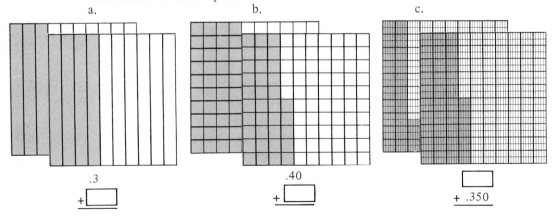

---
*T.P. Carpenter, M.K. Corbitt, H.S. Kepner, M.M. Lindquist, and R.E. Reys, "Decimals: Results and Implications from National Assessment," *The Arithmetic Teacher,* No. 4 (April 1981), pp. 34–37.
**M.J. Driscoll, "The Role of Manipulatives in Elementary School Mathematics," *Research Within Reach: Elementary School Mathematics,* (St. Louis, Missouri: Cemrel, Inc., 1983), p. 7.

★ d. State a general rule (using the vertical form) for adding two decimals.

2. **Subtraction:** Subtraction of decimals can be illustrated with squares by encircling the amount to be taken away. Compute the following differences by encircling the shaded amount to be taken away and counting the number of shaded parts that remain.

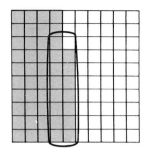

.47 − .15 = .32

a.   b.   c.

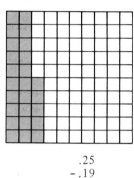 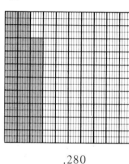 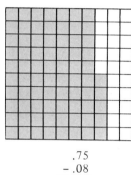

```
   .25           .280          .75
 − .19         − .175        − .08
```

d. State a general rule (using the vertical form) for subtracting one decimal from another.

3. **Multiplication by a Whole Number:** Here is the decimal square for .45. If there were 3 of these decimal squares, there would be a total of 3 × 45 = 135 shaded parts. Since there are 100 parts in each whole square, the total shaded amount would be 1 whole square and 35 parts out of 100.

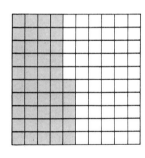

```
    .45
  ×   3
   1.35
```

a. If there were 6 decimal squares for .83, there would be a total of ☐ shaded parts. Since there are 100 parts in each whole square, this equals ☐ whole squares and ☐ parts out of 100.

```
    .83
  ×   6
```

132   CHAP. 7   DECIMALS: RATIONAL AND IRRATIONAL NUMBERS

★ b. If there were 4 decimal squares for .725, there would be a total of ☐ shaded parts. Since there are 1000 parts in each whole square, this equals ☐ whole squares and ☐ parts out of 1000.

.725
× 4
───
☐

c. State a general rule for multiplying a whole number times a decimal.

4. **Multiplying by Powers of 10:** Here is the decimal square for .72. If there were 10 of these decimal squares, there would be a total of 720 shaded parts. Since there are 100 parts in each whole square, the total shaded amount would be 7 whole squares and 20 parts out of 100.

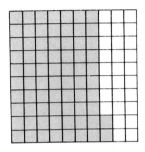

.72
× 10
────
7.20

★ a. If there were 10 decimal squares for .572, there would be a total of ☐ shaded parts. Since there are 1000 parts in each whole square, this equals ☐ whole squares and ☐ parts out of 1000.

.572
× 10
────
☐

b. If there were 100 decimal squares for .806, there would be a total of ☐ shaded parts. Since there are 1000 parts in each whole square, this equals ☐ whole squares and ☐ parts out of 1000.

.806
× 100
─────
☐

★ c. State a general rule for multiplying a decimal by a power of 10.

5. **Multiplication by a Decimal:** The product .3 × .1 means "take .3 of .1." To do this we can split the shaded amount of a .1 square into 10 equal parts and take 3 of them. Since each new part is one hundredth of a whole square, .3 times .1 is equal to .03.

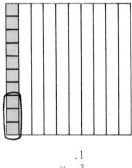

.1
× .3
────
.03

ACTIVITY SET 7.2   133

★ a. Use a decimal square for .2 to show why .3 × .2 = .06.

b. The product .08 × .1 means "take ☐ of ☐." To do this we can split the shaded amount of a ☐ square into ☐ equal parts and take ☐ of them. Since each new part is one thousandth of a whole square, .08 times .1 is equal to ☐.

```
   .1
 × .08
 ─────
   ☐
```

c. Use a decimal square for .2 to show why .08 × .2 = .016.

★ d. The product .06 × .01 means "take ☐ of ☐." To do this we can split the shaded amount of a ☐ square into ☐ equal parts and take ☐ of them. Since each new part is one ten-thousandth of a whole square, .06 times .01 is equal to ☐.

```
   .01
 × .06
 ─────
   ☐
```

e. State a general rule for multiplying a decimal times a decimal.

6. **Division by a Whole Number:** The *partitive concept of division* can be used to illustrate the division of a decimal by a whole number. The shaded amount of a square for .60 has been divided into 3 equal parts to illustrate .60 ÷ 3. Since each of these parts has 20 shaded parts, .60 divided by 3 is .20. Divide each decimal square into parts and use the result to determine the quotient.

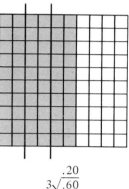

$3\overline{).60}^{\,.20}$

★ a.

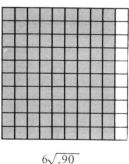

$6\overline{).90}$

b.

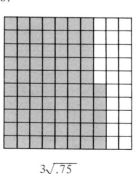

$3\overline{).75}$

c.

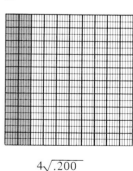

$4\overline{).200}$

d. Explain how decimal squares can be used to compute .1 ÷ 3 to three decimal places.

★ e. State a general rule for dividing a decimal by a whole number.

7. **Dividing by Powers of 10:** Use decimal squares and the partitive concept of division to illustrate each of the following quotients.

★ a. $.3 \div 10 = .03$ b. $.6 \div 100 = .006$ c. $.45 \div 10 = .045$

d. State a general rule for dividing a decimal by a power of 10.

8. **Division by a Decimal:** The *measurement concept of division* involves repeatedly measuring off or subtracting one amount from another. The square at the right shows that 15 of the shaded parts can be measured off from 75 parts 5 times.

$.15\overline{\smash{)}.75}$ with quotient 5

Draw lines to measure off the shaded parts of each decimal square and use the result to determine the quotient.

a.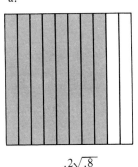

★ b.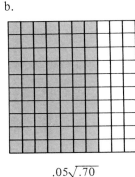

c.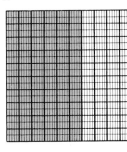

$.2\overline{\smash{)}.8}$ $.05\overline{\smash{)}.70}$ $.3\overline{\smash{)}.600}$

★ d. Explain how decimal squares can be used to show that $.9 \div .2 = 4.5$.

e. Students were asked to estimate $250 \div .5$ on the second mathematics assessment of the NAEP. Only 39% of the 17-year-olds correctly estimated this quotient, and 47% ignored the decimal point, giving an answer of 50.* Use the measurement concept of division to explain why $250 \div .5 = 500$.

f. State a general rule for dividing by a decimal.

*National Assessment of Educational Progress, "Math Achievement Is Plus and Minus," *NAEP Newsletter*, Vol. XII, No. 5 (October 1979), p. 2.

9. **All Four Operations** (Cut out the Decimal Squares from Material Cards 32, 33, 34, and 35 for Activities b and c):

   a. The following 10 decimals represent decimal squares:

   .3   .75   .8   .650   .15
   .2   .35   .45   .400   .350

   Complete these equations by placing the 10 decimals in the boxes.

   ☐ + ☐ = ☐
   ☐ − ☐ = ☐
   4 × ☐ = ☐
   ☐ ÷ ☐ = 3

   b. Spread the decimal squares face down and select 10 of them. Using the decimal from each square only once, complete as many of the four equations as possible. If you cannot complete all four equations, continue to select squares, one at a time, until all the equations can be completed.

   ☐ + ☐ = ☐
   ☐ − ☐ = ☐
   3 × ☐ = ☐
   ☐ ÷ ☐ = 2

   c. *Solitaire:* The activity in part **b** can be played like a solitaire game. See how many turns it takes you to complete the four equations by selecting only 10 decimal squares on each turn.

Just For Fun – **Decimal Games for Operations** (Material Cards 32, 33, 34, and 35)

Decimal Squares Black Jack (Addition, 2-4 players): The dealer shuffles the deck of decimal squares and gives one square face down to each player. The dealer also gets one square. The object is to use one or more squares to get a decimal or sum of two or more decimals which is less than or equal to 1, without "going over." Each time a player wants another square from the dealer, he/she says "hit me." After every player has said, "I'm holding," the players show their squares. The player whose sum is closest to 1 but not greater than 1, wins one point. The first player to win 5 points wins the game.

Greatest Quotient (Division, 2-4 players): Spread the decimal squares face down. Each player chooses two squares and computes the quotient of one decimal divided by the other. The object is to get the greatest quotient. *Chance Option:* If a player wishes to try to increase her/his quotient, then one more square may be chosen. However, before choosing a new square the player must discard one of her/his squares. The player with the greatest quotient wins all the squares used in that round, including any discarded squares. If there is a tie, the squares are placed aside and the winner of the next round gets them. Play continues until the deck has been played through and no more rounds are possible. The player who has won the most squares wins the game.

Greatest Difference (Subtraction, 2-4 players): This game is similar to the greatest quotient game except that the player with the greatest difference wins the round.

ACTIVITY SET 7.3

COMPUTING WITH THE CALCULATOR

In 1671 Gottfried Wilhelm Leibniz started the development of a mechanical computer. He wanted to mechanize the calculating in trigonometric and astronomical tables to free scientists from the drudgery of such work. "It is unworthy of excellent men," wrote Leibniz, "to lose hours like slaves in the labor of calculations." In the past few years we have been freed from the labor of elementary calculations by the availability of hand-held calculators. In the following activities you may wish to use a calculator for computing with percent and scientific notation. Percent will be used to calculate costs minus discounts, costs plus sales taxes, and interest rates that are compounded annually, quarterly, and monthly. We will also consider the quotients of whole numbers and the relationships of whole number remainders to the decimals obtained from computing with these quotients on a calculator.

1. Most calculators with a percent key are designed so that reductions in costs due to discounts can be conveniently computed. The five calculator steps and the displays that follow show how to find the cost of a $14.85 item with a 22% discount. Try these steps on your calculator to see if you get the same results.

Steps	Display
1. Enter 14.85	14.85
2. $\boxed{-}$	14.85
3. Enter 22	22.
4. $\boxed{\%}$	−3.267
5. $\boxed{=}$	11.583

    a. If your calculator does not have a percent key, the discounted cost of the previous item can be found by computing .22 × 14.85 and subtracting this from 14.85. Compute this product and difference.

    ★ b. The discounted cost in part **a** can be found in one step by computing .78 × 14.85. Compute this product. The reason this method works can be seen in the following equations. What number property is used in the first of these two equations?

    $$14.85 - (.22 \times 14.85) = (1 - .22) \times 14.85 = .78 \times 14.85$$

    c. Use your calculator to find the cost of a $35.89 skateboard with a discount of 18%. Use two of the three methods that have been shown to check your answer.

2. Most calculators with a percent key will compute the cost of an item plus the sales tax by five steps similar to those in Activity 1 for computing a discount. The following steps and displays show how to compute the cost of a $6.28 item with a 6% sales tax. Try these steps on a calculator with a percent key to see if you get the same displays.

Steps	Displays
1. Enter 6.28	6.28
2. [+]	6.28
3. Enter 6	6.
4. [%]	.3768
5. [=]	6.6568

a. The cost of the item in the preceding example can be found by computing .06 × 6.28 and adding it to 6.28. Compute this product and sum.

★ b. The cost of the item plus the tax in part a can be found in one step by computing 1.06 × 6.28. Use an equation to explain why .06 × 6.28 + 6.28 = 1.06 × 6.28.

c. Find the cost of a $12.89 calculator with a sales tax of 4%.

d. Find the cost of a $19.30 sweater with a 15% discount and a sales tax of 3%.

3. If your calculator adds on percentages such as by the steps in Activity 2, you will find it easy to compute compound interest on loans or savings. The next sequence of steps shows how to compute the yearly amount of savings for a $300 deposit at 6% interest compounded annually. The displays in Steps 2 and 3 show the amount of money in the account at the beginning of the 2nd and 3rd years.

a. Write in the amounts for the beginning of the 4th, 5th, and 6th years.

★ b. In how many years will the original deposit of $300 be doubled?

Steps	Displays	
1. Enter 300	300.	
2. [+] 6 [%] [=]	318.	(Beginning of 2nd year)
3. [+] 6 [%] [=]	337.08	(Beginning of 3rd year)
4. [+] 6 [%] [=]		(Beginning of 4th year)
5. [+] 6 [%] [=]		(Beginning of 5th year)
6. [+] 6 [%] [=]		(Beginning of 6th year)

4. When the interest in a savings account is compounded quarterly, each 3-month period the savings earns an interest that is one-fourth of the annual interest rate. If the interest is 6%, the quarterly rate is 1.5%. The amount of savings at the end of each quarter can be determined by replacing 6% by 1.5% in the steps in Activity 3.

a. Compute the savings account balances for Steps 3, 4, and 5 shown on the following page.

b. Compare the savings account balances at the beginning of the second year for interest compounded annually (Step 2, Activity 3), with interest compounded quarterly (Step 5 below). Which method produces the greater savings? How much more is saved?

c. How much will be in the savings account after 1 year if $300 is deposited at an annual interest rate of 6% compounded monthly? (*Hint:* The interest rate is .5% per month and is compounded 12 times a year.)

Steps	Displays	
1. Enter 300	300.	
2. [+] 1.5 [%] [=]	304.5	(Beginning of 2nd quarter)
3. [+] 1.5 [%] [=]		(Beginning of 3rd quarter)
4. [+] 1.5 [%] [=]		(Beginning of 4th quarter)
5. [+] 1.5 [%] [=]		(Beginning of 2nd year)

5. Some calculators have a special button such as [EE] or [EEX] for displaying numbers and computing with numbers in scientific notation. In some cases, a number will be displayed in scientific notation by entering the number and pressing [EE] and [=], as shown above. The scientific notation for this number is 7.493×10^8. The *mantissa,* a number from 1 to 10, appears on the left of the display and the *characteristic,* an exponent of 10, appears on the right. In other types of calculators, numbers will be displayed in scientific notation by entering the number into the calculator and pressing [=]. Does your calculator convert numbers into scientific notation?

Enter [7 4 9 3 0 0 0 0 0]
Press [EE] [7.4 9 3 _ _ _ 0 8]
      Mantissa   Characteristic

a. Each of the following displays shows a number that has been entered on the calculator. Write the number that will appear on the calculator if this number is displayed in scientific notation.

Enter [_ _ _ _ 4 7 0 0 0]   [_ _ 1 0 0 0 0 0 0]   [3 9 2 8 4 0 0 0 0]
Scientific Notation [_ _ _ _ _ _ _ _]   [_ _ _ _ _ _ _ _]   [_ _ _ _ _ _ _ _]

b. The scientific notation for a number less than 1 will have a negative characteristic. This is illustrated for the decimal .00384 shown below. Write the scientific notation for the other two numbers as they will appear on a calculator.

Enter [_ _ _ _ .0 0 3 8 4]   [_ _ _ _ .4 1 0 0 6 8]   [_ _ _ _ .0 0 0 0 8 7]
Scientific Notation [3.8 4 _ _ _ _ ⁻3]   [_ _ _ _ _ _ _ _]   [_ _ _ _ _ _ _ _]

c. What is the largest number that can be displayed in scientific notation if the calculator has eight places for the digits in the mantissa and two places for the characteristic? Write this number in scientific notation.

★ d. What is the smallest positive number that can be displayed on the calculator in part c? Write this number in scientific notation.

6. Calculators with scientific notation will automatically use this form whenever the numbers in positional numeration are too small or too large for the display. For example, 73200 × 639000 will appear as shown on the above display. Compute the following products and write the digits that will appear in the display of a calculator having scientific notation.

$$\underline{4.67748}\ \underline{1\,0}$$
Mantissa　　Characteristic

★ a. 1,000,000 × 1,000,000　　　[_____]

　b. 47,000,000 × 19,200　　　[_____]

★ c. .000008 × .00000034　　　[_____]

　d. $(6.18 \times 10^{15}) \times (2.3 \times 10^{40})$　　　[_____]

7. Some calculators compute with extra "hidden" digits which are not shown in their display. Assume that you have a calculator which displays 10 digits and keeps three more hidden. If you were to compute the decimal for 18/17 on this calculator, it would display 1.058823529, and the next three digits, namely, 411, would be kept internally. It is easy to find out if your calculator does this. In this example, subtract 1 from 1.058823529 and there will be room for one more digit in the display.

　a. Compute 18/17 on your calculator and subtract 1. Does the next digit in the decimal for 18/17 appear in the display?

　b. If you computed the decimal for 1/17 on the preceding calculator, the decimal .0588235294 would appear, and the next three digits, 117, would be kept internally. What operation can be performed in order to get rid of the "0" so that .5882352941 will appear?

★ c. Continuing part b, what operations can be performed on .5882352941 to get rid of the "5" so that .8823529411 will appear?

8. a. The fractions in the sequence

$$\frac{1}{2}\ \frac{1}{3}\ \frac{1}{4}\ \frac{1}{5}\ \frac{1}{6}\ \frac{1}{7}\ \frac{1}{8}\ \cdots$$

get closer and closer to zero but their sum can be made arbitrarily large. That is, by continuing in this sequence and adding more and more of these fractions, we can make the sum as large as we please. Use your calculator to complete table (a) by replacing each fraction with a decimal. If your calculator has a memory, it can be used to store each sum as you calculate the decimal for the next fraction.

★ b. The next sequence of fractions also gets closer and closer to zero.

$$\frac{1}{2}\ \frac{1}{4}\ \frac{1}{8}\ \frac{1}{16}\ \frac{1}{32}\ \frac{1}{64}\ \frac{1}{128}\ \cdots$$

Compute their sums in table (b). No matter how many of these fractions you add, their sum is always below a certain number. What is this number?

(a)

$\frac{1}{2} = .5$
$\frac{1}{2} + \frac{1}{3} = .8\overline{3}$
$\frac{1}{2} + \frac{1}{3} + \frac{1}{4} = 1.08\overline{3}$
$\frac{1}{2} + \frac{1}{3} + \frac{1}{4} + \frac{1}{5} =$
$\frac{1}{2} + \frac{1}{3} + \frac{1}{4} + \frac{1}{5} + \frac{1}{6} =$
$\frac{1}{2} + \frac{1}{3} + \frac{1}{4} + \frac{1}{5} + \frac{1}{6} + \frac{1}{7} =$
$\frac{1}{2} + \frac{1}{3} + \frac{1}{4} + \frac{1}{5} + \frac{1}{6} + \frac{1}{7} + \frac{1}{8} =$
$\frac{1}{2} + \frac{1}{3} + \frac{1}{4} + \frac{1}{5} + \frac{1}{6} + \frac{1}{7} + \frac{1}{8} + \frac{1}{9} =$
$\frac{1}{2} + \frac{1}{3} + \frac{1}{4} + \frac{1}{5} + \frac{1}{6} + \frac{1}{7} + \frac{1}{8} + \frac{1}{9} + \frac{1}{10} =$

(b)

$\frac{1}{2} = .5$
$\frac{1}{2} + \frac{1}{4} = .75$
$\frac{1}{2} + \frac{1}{4} + \frac{1}{8} =$
$\frac{1}{2} + \frac{1}{4} + \frac{1}{8} + \frac{1}{16} =$
$\frac{1}{2} + \frac{1}{4} + \frac{1}{8} + \frac{1}{16} + \frac{1}{32} =$
$\frac{1}{2} + \frac{1}{4} + \frac{1}{8} + \frac{1}{16} + \frac{1}{32} + \frac{1}{64} =$
$\frac{1}{2} + \frac{1}{4} + \frac{1}{8} + \frac{1}{16} + \frac{1}{32} + \frac{1}{64} + \frac{1}{128} =$
$\frac{1}{2} + \frac{1}{4} + \frac{1}{8} + \frac{1}{16} + \frac{1}{32} + \frac{1}{64} + \frac{1}{128} + \frac{1}{256} =$
$\frac{1}{2} + \frac{1}{4} + \frac{1}{8} + \frac{1}{16} + \frac{1}{32} + \frac{1}{64} + \frac{1}{128} + \frac{1}{256} + \frac{1}{512} =$

9. On this calculator there are 8 places for digits, and the decimal point may occur at the left or right of the display or between any of the digits. The number 72.833333 was obtained by dividing 1748 by 24. This tells us that the quotient is greater than 72 but it does not give the whole number remainder. Since 72 × 24 = 1728, the remainder when dividing 1748 by 24 is 20.

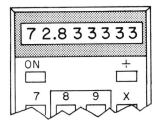

a. Enter .8333333 onto your calculator and multiply by 24. How can this product be used to get the remainder of 20 in the example of 1748 ÷ 24?

★ b. Use your calculator to compute 1472081 ÷ 3876. Find the whole number remainder that is represented by the decimal part of your answer. Explain how this can be done by using a calculator.

c. The product 5268000 × 390000 will exceed the capacity of a calculator with 8 places for digits. Try it. Explain how you could use such a calculator if it did not have scientific notation to compute this product. (*Hint:* Use scientific notation.)

Challenge: The decimal for 1/19 has 18 digits before it repeats the pattern. By dividing 1 by 19, you can get the first 7 or 8 digits on your calculator. Find a way to obtain the first 18 digits by using a calculator. Explain how this can be done.

Just for Fun—Number Search

Find equal pairs of decimals, percents, and fractions. See if you can find 20 such pairs.

ACTIVITY SET 7.4

IRRATIONAL NUMBERS ON THE GEOBOARD

Does there exist a number (whole number, fraction, or irrational number) which will divide into the lengths of the side and diagonal of a square a whole number of times? In this square, for example, it looks as though there is a small unit that will divide into both *AB* and *AC* a whole number of times. Will there be such a unit for the side and hypotenuse of any square? How does your intuition tell you to answer this question?

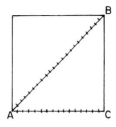

About 2300 years ago mathematicians would have answered yes! They felt that any two line segments are *commensurable,* that is, have a common unit of measure. This unit might be very small, as a millionth or a billionth of a centimeter, but surely, there would always be such a length. When the Greek mathematician Hippasus (470 B.C.) discovered that the side and hypotenuse of a square are *incommensurable*—that is, do not have a common unit of measure—it caused such a crisis in mathematics that, reportedly, he was taken to sea and never returned. His discovery ultimately led to new types of numbers, called *irrational numbers,* which are infinite nonrepeating decimals.*

The geoboard and dot paper will be used in the following activities to form figures whose sides and perimeters have lengths that are irrational numbers. The areas of squares, triangles, and rectangles will be determined by computing with irrational numbers and checking the results by techniques for finding areas on the geoboard. The Pythagorean theorem will be verified for several right triangles by finding the areas of squares on dot paper.

1. There are eight different (noncongruent) squares that can be formed on the geoboard. One of these is shown below. Sketch the remaining squares on these geoboards.

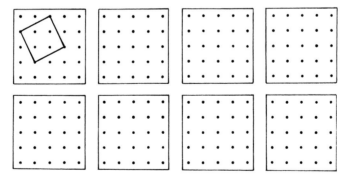

a. Under each square write its area in square units. To find some of these areas, it will be helpful to enclose the square in a larger square. The square on the first geoboard can be enclosed in a 3 by 3 square of area 9 square units. The area of the smaller square can be found by subtracting off the outside area.

---

*For a proof that the side and hypotenuse of a square are incommensurable, see R. Courant and H. Robbins, *What Is Mathematics* (New York: Oxford University Press, 1941), pp. 59–60.

★ b. The length of the side of a square of area A is $\sqrt{A}$. For example, one of our eight squares has an area of 10 square units and so the length of its side is $\sqrt{10}$. Write the areas of these eight squares and the length of their sides in the table. List the areas in order from least to greatest.

Area of Square	1	2						
Length of Side	1	$\sqrt{2}$						

2. On the following grid there are four right triangles with A and B squares on each of their legs. The C square is on the hypotenuse of each triangle.

★ a. Find the area of each A and B square and write it inside the square. The areas of the squares for the two lower right triangles can be found by enclosing them in larger squares.

★ b. Find the area of each C square by enclosing it in a larger square and subtracting off the outside area.

c. By the Pythagorean theorem, the area of square A plus the area of square B is equal to the area of square C. Use this relationship to check your answers in parts a and b.*

d. If a square has an area A, its side has length $\sqrt{A}$. Use the areas of these squares to write the lengths of the sides of each triangle.

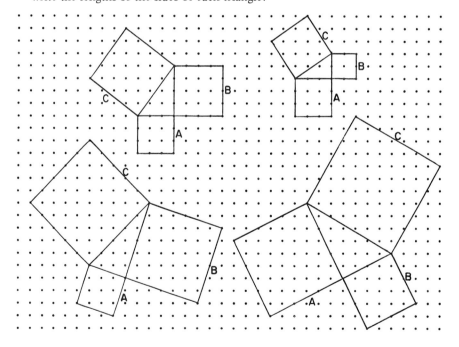

*For another demonstration of the Pythagorean theorem, see A.L. Buchman, "An experimental approach to the Pythagorean theorem," *The Arithmetic Teacher,* **17** No. 2 (February 1970), pp. 129-32.

3. Here are 10 of the 14 line segments of different lengths which can be formed on the geoboard. Each of these segments is the hypotenuse of a right triangle. (The *hypotenuse* is the side of the triangle which is opposite the right angle.) For example, the line segment whose length is $\sqrt{13}$ is the hypotenuse of a right triangle whose legs have lengths 2 and 3. Sketch a right triangle for each of these segments and use the Pythagorean theorem to determine the length of each segment.

★ a.   b.   ★ c.   d.   e.

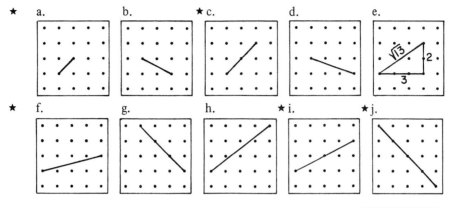

★ f.   g.   h.   ★ i.   ★ j.

4. Sketch line segments on the geoboards below for each of the given lengths. The line segment on the geoboard at the right is 3 times longer than $\sqrt{2}$. This is written as $3\sqrt{2}$.

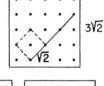

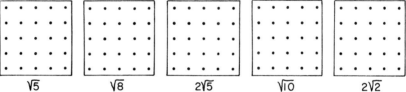

$\sqrt{5}$ $\sqrt{8}$ $2\sqrt{5}$ $\sqrt{10}$ $2\sqrt{2}$

Compare the lengths of these line segments and write $=$, $>$, or $<$ between the following pairs of numbers.

★ a. $2\sqrt{2}$ $\sqrt{5}$ ★ c. $\sqrt{2}+\sqrt{8}$ $\sqrt{10}$ ★ e. $3\sqrt{2}$ $2\sqrt{5}$

b. $3\sqrt{2}$ $\sqrt{10}$ d. $2\sqrt{2}$ $\sqrt{8}$ f. $\sqrt{10}$ $2\sqrt{5}$

g. One of the following equations is true for all positive real numbers r and s, and the other two are false. Use the inequalities in parts **a** through **f** to determine the two that are false. Indicate which counterexamples you are using.

$$\sqrt{r}+\sqrt{s}=\sqrt{r+s} \qquad \sqrt{r}\times\sqrt{s}=\sqrt{r\times s} \qquad \sqrt{r\times s}=r\sqrt{s}$$

★ 5. This hexagon has a perimeter of $6 + \sqrt{2} + \sqrt{8}$ centimeters. If we use a centimeter ruler to approximate $\sqrt{2}$ and $\sqrt{8}$, this perimeter will be approximately equal to $6 + 1.4 + 2.8 = 10.2$ centimeters.

Use the results in Activity 3 to find the lengths of the sides of each polygon below. Write the perimeters of these polygons in the table. Then use your centimeter ruler to get decimal approximations for these perimeters.

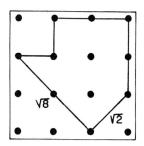

Polygon	Example	a	b	c	d	e	f
Perimeter	$6+\sqrt{2}+\sqrt{8}$						
Approximate Perimeter	10.2						

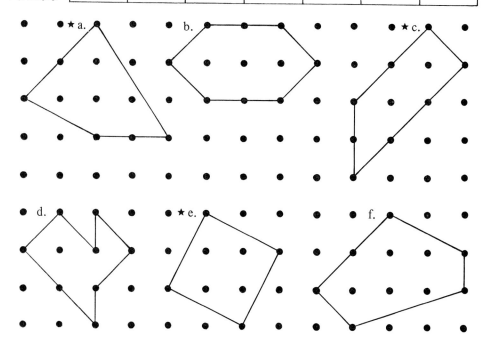

6. The formulas for the area of a rectangle, length times width, and the area of a triangle, 1/2 times base times altitude, continue to hold when the lengths of the sides of these polygons are irrational numbers. Use these formulas and the fact that $\sqrt{a} \times \sqrt{b} = \sqrt{ab}$ to determine the areas of the following figures. Check your answers by counting squares and half-squares, or by enclosing the polygon in a square or rectangle and subtracting the outer area.

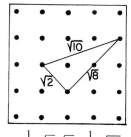

Area $= \frac{1}{2}\sqrt{2}\sqrt{8} = \frac{1}{2}\sqrt{16} = 2$

146 CHAP. 7 DECIMALS: RATIONAL AND IRRATIONAL NUMBERS

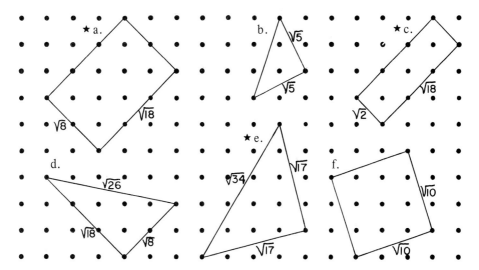

★ 7. **Challenge:** There are two line segments of different lengths on a 2 by 2 array and five on a 3 by 3 array. How many such line segments can be formed on 4 by 4 and 5 by 5 arrays? You will find a pattern in the number of line segments for the first four arrays, which suggests that there should be 20 line segments of different length on a 6 by 6 array. Find the number of line segments of different length on a 6 by 6 array and explain why there are not 20 such lines.

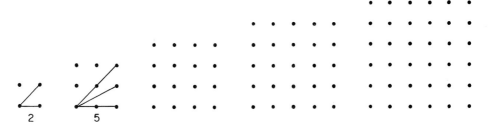

Just for Fun--Golden Rectangles

Some rectangles are more aesthetically pleasing than others. One in particular, the golden rectangle, has been a favorite of architects, sculptors, and artists for over 2000 years. The Greeks of the fifth century B.C. were fond of this rectangle. The front of the Parthenon in Athens fits into a golden rectangle. In the nineteenth century the German psychologist Gustav Fechner found that most people do unconsciously favor golden rectangles. Which one of the rectangles shown next do you feel has the most pleasing dimensions?

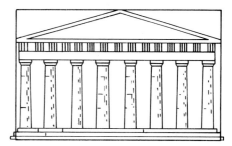

ACTIVITY SET 7.4   147

The length of a *golden rectangle* divided by the width is an irrational number which when rounded to three decimal places is 1.618. Measure the preceding rectangles to determine which is closest to a golden rectangle. Measure some other rectangles such as greeting cards, index cards, pictures, mirrors, and so on, to see if you can find a golden rectangle. (A ratio of 1.6 is close enough.) Check the ratio of the length to the width of the front of the Parthenon.

A golden rectangle can be constructed from a square such as *ABCD* by placing a compass on the midpoint of $\overline{AB}$ and swinging an arc *DF*. Then *FBCE* is a golden rectangle. Make a golden rectangle and check the ratio of its sides. If the sides of the square have length 2, then $BF = 1 + \sqrt{5}$. What is $(1 + \sqrt{5})/2$ to 3 decimal places? Cut out your rectangle and fold it in half along its longer side. Is the result another golden rectangle?

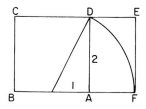

The ratio of the sides of a golden rectangle is called the *golden mean* or *golden ratio*.* It is surprising that this number (approximately 1.618) is also associated with the Fibonacci numbers: 1, 1, 2, 3, 5, 8, 13, ... The ratios of successive terms in this sequence get closer and closer to the golden ratio. The ratio of 8/5, for example, is 1.6. How far must you go in the Fibonacci sequence before the ratio of numbers is approximately 1.618?

*Several occurrences of the golden ratio are described by C.F. Linn, *The Golden Mean* (New York: Doubleday, 1974).

8

Geometry and Algebra

ACTIVITY SET 8.1

GEOMETRIC MODELS FOR ALGEBRAIC EXPRESSIONS

The ancient Greek mathematicians at the time of Pythagoras represented the product of two numbers geometrically as a square or a rectangle. A number times itself, such as 6 times 6, was represented by a square with sides of length 6. Because of this practice, 6 times 6 was called 6 *squared* and the product, 36, which is the area of the square, was called a *square number*. Similarly, the product of any two numbers was represented by a rectangle whose length and width are the two numbers. Four times seven, for example, was pictured as a 4 by 7 rectangle, and the product, 28, which is the area of the rectangle, was called a *rectangular number*.

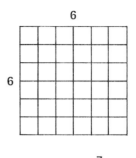

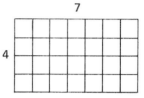

In this activity set the products of algebraic terms will be represented geometrically as areas of squares and rectangles. Copies of the following geometric pieces are on Material Card 6 and should be cut out for the activities.

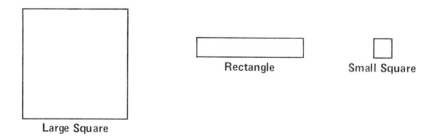

In Activities 1 through 3 we will let the geometric pieces have the following dimensions, where x and y are arbitrary positive numbers with $y < x$.

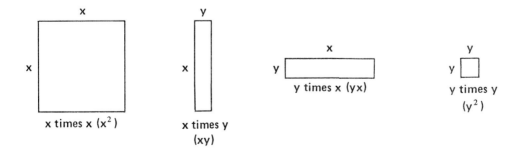

150   CHAP. 8   GEOMETRY WITH COORDINATES

1. The product $x + y$ times $x + y$ is illustrated by the square whose sides have length $x + y$. This square has four parts.

   a. Label each part of the square with the algebraic term it represents.

   ★ b. Use this illustration and the fact that $xy = yx$ to write the following product as the sum of three algebraic terms.

   $$(x + y)(x + y) =$$

2. Here are four more geometric models for the products of algebraic terms. Label each part of the model with the product it represents, and then use the model to obtain the product of the algebraic terms.

   ★ a.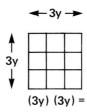

   $$(3y)(3y) =$$

   b.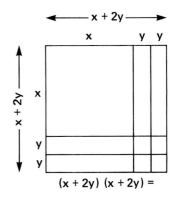

   $$(x + 2y)(x + 2y) =$$

   ★ c.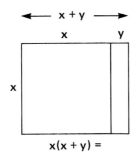

   $$x(x + y) =$$

   d.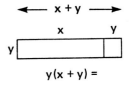

   $$y(x + y) =$$

3. Build a model with the geometric pieces to represent each product. Sketch the geometric model and label each part. Use the model to obtain the product.

   ★ a. $y(2x + y) =$

ACTIVITY SET 8.1 151

b. $x(x + 2y) =$

c. $(x + 3y)^2 =$

d. $(2y)^2 =$

In Activities 4 through 7 we will let the geometric pieces have the dimensions shown below. The small 1 by 1 square will be called the *unit square*.

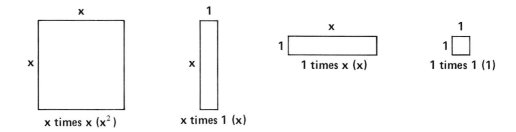

4. The figure at the right is a rectangle with 1 large square, 5 rectangles, and 6 unit squares. Try using the following geometric pieces to form a rectangle. Sketch the rectangle if it can be formed. Which sets of pieces cannot be used to form rectangles?

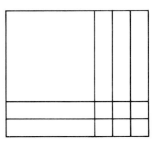

★ a. 1 square
   3 rectangles
   1 unit

b. 1 square
   3 rectangles
   2 units

★ c. 1 square
   7 rectangles
   12 units

d. 1 square
   5 rectangles
   3 units

e. 1 square
   7 rectangles
   10 units

152 CHAP. 8 GEOMETRY AND ALGEBRA

5. The figure shown here has 2 large squares, 7 rectangles, and 3 unit squares. Which of the following sets of geometric pieces can be used to form a rectangle? Sketch the rectangle if it can be formed.

★ a. 2 squares
   6 rectangles
   4 units

b. 2 squares
   9 rectangles
   4 units

★ c. 2 squares
   3 rectangles
   5 units

d. 2 squares
   9 rectangles
   10 units

e. 2 squares
   5 rectangles
   2 units

6. There are two ways of expressing the area of this figure. First, it has 1 large square, 5 rectangles, and 6 unit squares, so the area is $x^2 + 5x + 6$. Second, the width is $x + 2$ and the length is $x + 3$, so the area is the product of two factors: $(x + 2)(x + 3)$. Therefore $x^2 + 5x + 6 = (x + 2)(x + 3)$. Use the width and length of each figure to find the factors of the corresponding algebraic expression.

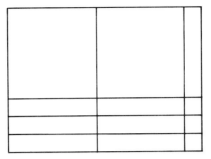

★ a.

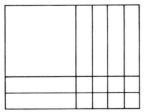

$x^2 + 6x + 8 =$
(   )(   )

b.

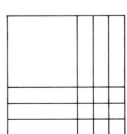

$x^2 + 6x + 9 =$
(   )(   )

c.

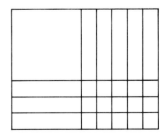

$x^2 + 8x + 15 =$
(   )(   )

★ d.   e.

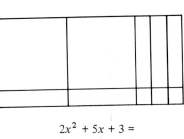   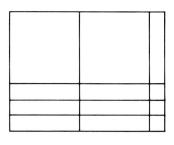

$2x^2 + 5x + 3 =$ $2x^2 + 7x + 3 =$
(   )(   )   (   )(   )

7. Use the geometric pieces listed below to try forming squares or rectangles. Two cannot be done. If a square or rectangle can be formed, write the factors of the algebraic expression.

★ a. 1 square, 7 rectangles, 10 units   b. 3 squares, 5 rectangles, 2 units
     $x^2 + 7x + 10$ $3x^2 + 5x + 2$

c. 1 square, 6 rectangles, 9 units   ★ d. 1 square, 1 rectangle, 1 unit
   $x^2 + 6x + 9$ $x^2 + x + 1$

e. 2 squares, 5 rectangles, 3 units   f. 3 squares, 7 rectangles, 4 units
   $2x^2 + 5x + 3$ $3x^2 + 7x + 4$

g. 2 squares, 7 rectangles, 6 units   h. 1 square, 4 rectangles, 4 units
   $2x^2 + 7x + 6$ $x^2 + 4x + 4$

i. 1 square, 7 rectangles, 9 units
   $x^2 + 7x + 9$

Just for Fun—Algebraic Skill Game*

This is a game in which algebraic expressions are matched to word descriptions. There are copies of algebraic skill cards on Material Card 22, which should be cut out for this game. There are two decks of 12 skill cards, one for each team. Each student has a skill card similar to the samples at the right.

Student #1

| **I have** |
| $n + 1$ |
| **Who has** |
| two less than a number |

Student #2

| **I have** |
| $n - 2$ |
| **Who has** |
| one more than two times a number |

Play: The game begins with student #1 saying, "I have $n + 1$. Who has two less than a number?" Then student #2 says, "I have $n - 2$. Who has one more than two times a number?" The game continues in this manner.

Each deck of 12 cards is circular. That is, the last card in the deck calls the first card. Therefore, all the cards in a deck should be used. If there are fewer than 12 students on a team, some students can use more than one card. If there are more than 12 students on a team, 2 students can share a card, or more cards can be made.

Objective: The two teams compete against each other for the shortest time required to complete the cycle of cards.

Variation: The game can be changed so that word descriptions are matched to algebraic expressions. Two sample cards are shown here. In this case student #1 says, "I have two more than three times a number. Who has $4x - 1$?" Student #2 says, "I have one less than four times a number. Who has $3x + 5$?" Make a deck of these cards and play this version of the game.

| **I have** |
| two more than three times a number |
| **Who has** |
| $4x - 1$ |

| **I have** |
| one less than four times a number |
| **Who has** |
| $3x + 5$ |

*T. Giambrone, "I HAVE . . . WHO HAS . . . ?" *The Mathematics Teacher,* 73, No. 6 (October 1980), pp. 504–06.

McDonnell Planetarium in St. Louis, Missouri

ACTIVITY SET 8.2

CONIC SECTIONS

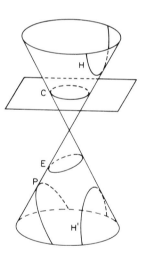

The Greeks studied the mathematical curves called *conic sections*. These curves can be obtained by the intersections of planes with a double cone. Depending on the angle at which the plane intersects the cone, the intersection may be a circle, ellipse, parabola, or hyperbola. In this figure, C is a *circle,* E is an *ellipse,* P is a *parabola,* and H and H' are the two branches of a *hyperbola*. It is somewhat of a mystery why the Greeks selected these curves to study, but from the present standpoint, it is exceedingly lucky. The conic sections paved the way for modern astronomy. In the seventeenth century these curves became important for describing the paths of planets and comets. If a comet is in orbit about the sun, its path is elliptical. If a comet enters the solar system and then leaves again, its path is parabolic or hyperbolic. The conic sections are currently used in the design of cables, buildings, roadways, bridges, and lenses for glasses, telescopes, and microscopes. The McDonnell Planetarium in St. Louis has a hyperbolic shape. This was chosen by designer Gyo Obata, because the hyperbolic paths of certain comets suggest, as he said, "the drama and excitement of space exploration."

1. **Drawing an Ellipse:** There is graph paper with a rectangular coordinate system on Material Card 9. Tack or tape this sheet onto a board and place two tacks at points (⁻4,0) and (4,0). Label them F_1 and F_2. Cut a piece of string and tie the ends so that you have a loop which has a circumference of 26 centimeters. Place the string around the tacks, and keeping tension with a pencil, draw an ellipse.

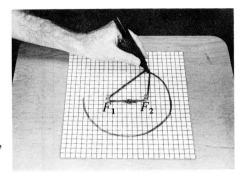

★ a. An *ellipse* is the set of points in a plane such that the sum of the distances from each point to two fixed points, called *focuses*, is a constant. As the ellipse is being drawn the sum of the distances from the pencil to focuses F_1 and F_2 is the same at all times. What is this sum?

b. The *major axis* of this ellipse is the portion of the x-axis which is inside the ellipse, and the *minor axis* is the portion of the y-axis inside the ellipse. What are the lengths of the major and minor axes?

★ c. Move the focuses to (⁻2,0) and (2,0) and use the same loop of string to draw an ellipse. What happens to the shape of the ellipse? What are the approximate lengths of the major and minor axes?

d. Place F_1 and F_2 both at (0,0) with one tack, and using the same loop of string and method as before, draw a curve. What type of curve do you get?

2. **Drawing a Hyperbola:*** Tape the rectangular coordinate system on Material Card 10 onto a board. Place two tacks for focuses F_1 and F_2 at (⁻5,0) and (5,0), and label point P at (6,7). Loop a piece of string around a pencil at point P and about F_1 and F_2 as shown in this picture. Hold both ends of the string below F_1 with your left hand. Hold the string against the pencil lead with your right forefinger (or tie a knot) so the pencil will not slip along the string. By pulling the two ends of the string in your left hand, the pencil will trace out the upper half of one branch of a hyperbola. When the pencil gets to $\overline{F_1F_2}$, hold the pencil against the lower string and continue tracing the curve.

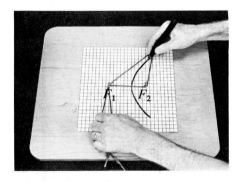

---

*Several more methods of creating hyperbolas are described by M. Gardner, "On conic sections, ruled surfaces and other manifestations of the hyperbola," *Scientific American,* **237** No. 3 (September 1977), pp. 24-42.

a. A *hyperbola* is the set of all points in the plane such that the difference between the distances from any point to two fixed points, called *focuses,* is a constant. Measure the distances from P to the focuses F_1 and F_2. What is the difference between these distances?

★ b. As the left hand pulls the string, what happens to the distances from the pencil to F_1 and F_2? What is the difference between each of these distances?

★ c. Explain why this difference stays the same as the string is pulled with the left hand.

d. Fold the paper about the y-axis and trace the other branch of the hyperbola. Select any point K on this branch and compute $KF_2 - KF_1$. What should this distance be?

3. **Paper-folding Parabolas:** Turn a piece of paper sideways and place a point 4 centimeters above the lower edge of the paper. Label it F. Fold a corner of the paper so that the lower edge touches point F, as shown in this figure. Open the paper and draw a line along this crease. This can be easily done without a ruler by partially opening the folded paper and running your pen or pencil along the crease. Repeat this activity many times by folding different points on the lower edge of the paper up to point F. The crease lines form what is called the *envelope of a parabola.* Sketch in this curve.

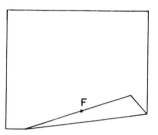

★ a. A *parabola* is the set of points in a plane which are the same distance from a fixed point called the *focus* as from a fixed line called the *directrix.* The paper-folded parabola has focus F. Where is the directrix of this parabola?

b. Place a point 10 centimeters above the lower edge of the paper and repeat the folding activity. How does the shape of the parabola differ from that of the parabola you got by using point F?

c. What will happen to the shape of the parabola if point F is placed one-half of a centimeter above the lower edge of the paper?

4. **Paper-folding Ellipses:** Draw a circle of radius 8 centimeters (compass on Material Card 41). Label a point P which is 5 centimeters from the center, C. Fold a point on the circumference of the circle onto point P and draw a line on the crease (see figure). Repeat this activity by folding about 20 different points from around the circle onto point P. The crease lines you get will be the *envelope of an ellipse*. Sketch in this curve.

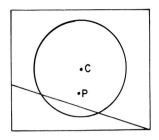

★ a. Point P is one focus of the ellipse. What is the other focus?

b. Pick any point on the ellipse and find the sum of the distances to the two focuses (points P and C). This sum should be approximately the same for any point on the ellipse. What is this sum?

★ c. What will happen to the shape of the ellipse as P moves closer to the center of the circle?

5. **Paper-folding Hyperbolas:** Draw a circle of radius 8 centimeters and label a point P outside the circle. Fold point P onto the circumference of the circle and draw a line on the crease (see figure.) Repeat this activity, folding the point P onto many different points around the circle. The lines you get from the creases are called the *envelope of a hyperbola*. Sketch in the two branches of this curve.

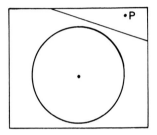

★ a. Point P is the focus of one branch of this hyperbola. Describe the location of the second focus.

b. What will happen to the shape of the hyperbola if point P is moved further from the circle?

ACTIVITY SET 8.2

★ 6. **Mystery Curve:** Point B is the center of a circle of radius 1.0 centimeter, and point A is the center of a circle with radius 2.0 centimeters. Both X and Y, which are the intersections of these two circles, are twice as far from point A as from point B. Find some more points that are twice as far from A as from B. (*Hint:* You may find it helpful to locate these points by drawing circles about A and B, with the radius of A twice the radius of B.) These points form one of the conic sections. Which one?

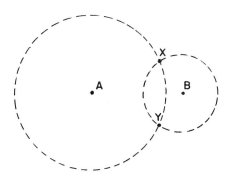

★ 7. **Spinning Cube Experiment:** Spin a cube, such as a die, on one of its corners. This can be done by holding the cube between the thumb on one hand and the forefinger on the other hand. Viewed from above, the spinning cube looks like the cone of a toy top. Viewed horizontally with your eyes on the same level as the cube, its spinning sides form one of the conic sections. Which one?

8. **Curves from Straight Lines:** Cut out a narrow strip of posterboard having a length of 10 centimeters and punch a small hole at its midpoint. Place this on two perpendicular lines so that each end touches one of the lines. Place a pencil at the midpoint, M, and with the help of a classmate, move the strip so that its end points always touch the two perpendicular lines. What curve is traced by M?

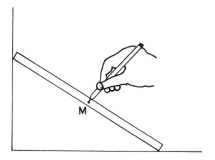

9. **Shadow Experiment:** A flashlight and a sphere, such as a basketball or a beachball, can be used in a darkened room to form shadows which are conic sections. Determine the shape of the shadow (circular, elliptical, parabolic, or hyperbolic) for each location of the light in the figures below. (*Hint:* Think of the rays of light as forming a cone and the surface the ball rests on as a plane which cuts the cone.)

★ a. b.

c. ★ d.

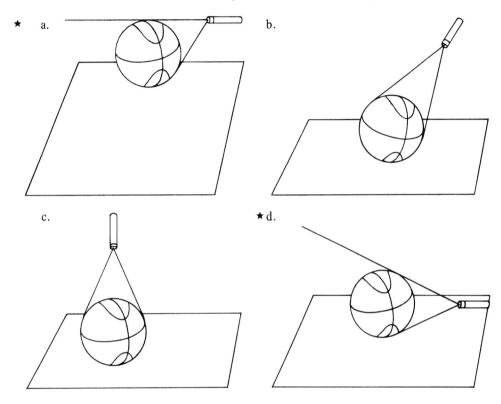

Just for Fun—Line Designs and Coordinate Games

1. **Line Designs:** *Line designs* are geometric patterns formed by connecting sequences of points with line segments. One common method is to connect equally spaced points on the sides of an angle. These points are connected in the following order: *A* to *a*, *B* to *b*, *C* to *c*, etc. The resulting design gives the illusion of a curve. The curve is a parabola and the line segments are the envelope of the parabola.*

    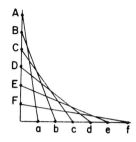

    Experiment with these different types of angles. For an obtuse angle the "curve" will be closer to the angle than for an acute angle.

    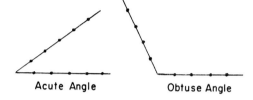

    Here are some art forms based on these parabolic angle patterns. The first one is formed by rubber bands and nails, the second by yarn and nails, and the third by white thread stitched onto black silk.**

2. **Battleships:*** This is a game for two players (or teams). Both players use a pair of rectangular grids such as the ones shown below (Material Card 4). Each player draws three battleships on one grid and records shots at the other player's battleships on the other grid. A battleship consists of three adjacent intersection points of the grid (from rows, columns, or diagonals) whose coordinates are whole numbers. Three examples are given.

    The players should not be able to see each other's grids. They take turns firing volleys of three shots at a time by giving the coordinates of each point. When hits are made, they are acknowledged immediately. A game ends when a player has lost all three battleships.

---

*For a proof of this fact, see K.P. Goldberg, "Curve Stitching in an Elementary Calculus Course," *The Mathematics Teacher,* **69** No. 1 (1976), pp. 12-14.

**See D.G. Seymour, L. Silvey, and J. Snider, *Line Designs* (Palo Alto: Creative Publications, 1968), for more examples of line designs.

***W.R. Bell, "Cartesian Coordinates and Battleships," *The Arithmetic Teacher,* **21** No. 5 (May 1974), pp. 421-22.

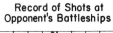

Three Battleships

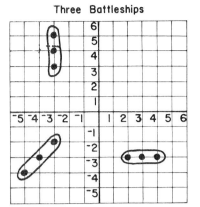

Record of Shots at Opponent's Battleships

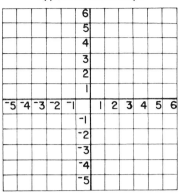

Variation: To speed-up the game you may wish to agree to inform your opponent whenever a shot is within one unit of a battleship.

3. **Hide-a-region:*** This game is played by two players (or teams) and is similar to Battleships. Both players use a pair of rectangular grids (Material Card 4). The game begins with each player sketching a rectangle on his/her grid with the coordinates of the vertices being integers. Both rectangles should have the *same area* but the players should not see each other's grids. The players take turns guessing points by giving their coordinates. Only one guess is allowed on each turn. Each player must tell whether the opponent's guess is an exterior, interior, vertex, or boundary point. Both players keep a record of their opponent's guesses and their own guesses. The first player to locate all four vertices is the winner.

Rectangle of Area 12

Record of Guesses to Locate Opponent's Rectangle

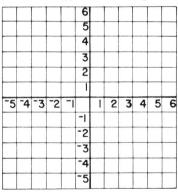

Variation: The players may use any type of regions as long as both have the same area.

---

*J.S. Overholser, "Hide-a-region, N ≥ 2 can play," *The Arithmetic Teacher,* **16** No. 6 (October 1969), pp. 496–97.

ACTIVITY SET 8.3

GEOMETRIC PATTERNS AND NUMBER SEQUENCES

Each tower in the above photo changes from the preceding one by some consistent geometric pattern. The numbers of blocks in these towers are the beginning of the sequence

$$1, 6, 15, 28, 45, \ldots$$

If the geometric pattern continues, what is the 10th number in this sequence? In this activity set you will use geometric patterns and number patterns to obtain a formula for the general term of a sequence.

1. **Tower Pattern:** The towers in the above photo can be constructed by the following method: the second tower can be obtained by putting 5 blocks under the first tower; the third tower can be obtained by putting 9 blocks under the second tower; the fourth tower can be obtained by putting 13 blocks under the third tower; etc. These observations lead to the following number pattern.

Tower 1	Tower 2	Tower 3	Tower 4
1	1 + 5	1 + 5 + 9	1 + 5 + 9 + 13

   You may also have noticed that the second tower has 2 blocks in the middle with 4 blocks around; the third tower has 3 blocks in the middle with 4 x (1 + 2) blocks around; the fourth tower has 4 blocks in the middle with 4 x (1 + 2 + 3) blocks around. This approach to constructing the towers leads to the following pattern.

Tower 1	Tower 2	Tower 3	Tower 4
1	2 + 4(1)	3 + 4(1 + 2)	4 + 4(1 + 2 + 3)

164 CHAP. 8 GEOMETRY AND ALGEBRA

a. Extend both of the preceding patterns to obtain the number of blocks in the tenth tower.

★ b. Use the second of the above patterns to write a formula for determining the number of blocks in the Nth tower.

c. Use the formula for the sum of consecutive whole numbers (see Activity Set. 1.1) to simplify the formula in part **b**. Check this formula by letting $N = 10$.

2. **Staircase Pattern**:

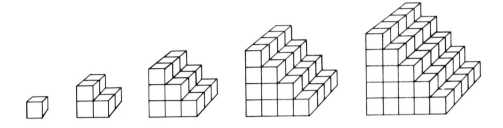

a. Find the number of blocks in each staircase. How many blocks are needed for the sixth staircase?

b. Find a pattern for expressing the numbers of blocks in these staircases.

★ c. Use your pattern from part **b** to predict the number of blocks in the tenth staircase.

d. One student noticed that the second staircase was 3/4 of a 2 by 2 by 2 cube; the third staircase was 4/6 of a 3 by 3 by 3 cube; and the fourth staircase was 5/8 of a 4 by 4 by 4 cube. Extend this pattern for the fifth and sixth staircases.

Staircase 1 $\frac{2}{2}$ of 1 by 1 by 1 $\frac{2}{2} \times 1 = 1$

Staircase 2 $\frac{3}{4}$ of 2 by 2 by 2 $\frac{3}{4} \times 8 = 6$

Staircase 3 $\frac{4}{6}$ of 3 by 3 by 3 $\frac{4}{6} \times 27 = 18$

Staircase 4 $\frac{5}{8}$ of 4 by 4 by 4 $\frac{5}{8} \times 64 = 40$

Staircase 5

Staircase 6

★ e. Write a formula for the number of blocks in the Nth staircase using the pattern from part **d**.

3. **U-Shaped Pattern**:

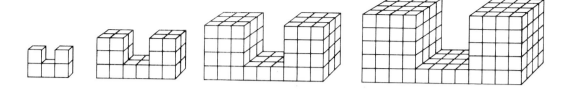

a. How many blocks are there in each U-shaped figure? How many blocks are needed for the fifth figure?

★ b. Find a pattern for expressing the numbers of blocks in these figures.

c. Use your pattern in part **b** to predict the number of blocks in the tenth figure.

★ d. Perhaps you noticed that each U-shaped figure can be obtained by removing a cube of blocks from the center of a rectangular solid (parallelepiped). This observation leads to the following number pattern. Continue this pattern for the fourth and fifth figures.

Figure 1 $1 \times 2 \times 3 - 1^3 = 5$

Figure 2 $2 \times 3 \times 6 - 2^3 = 28$

Figure 3 $3 \times 4 \times 9 - 3^3 = 81$

Figure 4

Figure 5

e. Write a formula for the number of blocks in the Nth figure using the pattern from part **d**.

4. Three-dimensional geometric patterns like those in Activities 1 through 3 are easy to produce. Create your own geometric figures by using blocks. Find a pattern for expressing the numbers of blocks in these figures. Use your number pattern to find a formula for the number of blocks in the Nth figure.

5. **Step Pattern:** Here are the first five figures in a two-dimensional pattern of squares and sums for the number of squares in each figure.

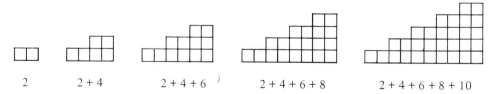

2   2 + 4   2 + 4 + 6   2 + 4 + 6 + 8   2 + 4 + 6 + 8 + 10

a. Explain how these sums were obtained from the figures.
b. What is the number of squares in the tenth figure?
c. Write a formula for the number of squares in the Nth figure.
★ d. Use the formula in part **c** to find the number of squares in the fiftieth figure.

6. **Dragon Pattern:**

a. How many squares are there in each figure? How many squares are needed for the sixth figure?
★ b. Find a pattern for expressing the numbers of squares in these figures.

Figure 1   Figure 2   Figure 3   Figure 4   Figure 5

c. Use your pattern in part **b** to determine the number of squares in the tenth figure.
d. Write a formula for the number of squares in the Nth figure. Use this formula to check your answer in part **c**.

7. Create a two-dimensional geometric pattern by using squares or by drawing on graph paper. Find a pattern for expressing the numbers of squares in these figures. Use your number pattern to find a formula for the number of blocks in the Nth figure.

Just for Fun—What's My Rule?

This is a function game that can be played by two players or teams. Team A agrees on a rule, such as, "double the number and add 1," and Team B tries to guess the rule. To obtain information about the rule, Team B selects a number (the independent variable x) and Team A uses their rule on this number to get a second number (the dependent variable y). Team B selects as many numbers as necessary to guess the rule. Then the teams change roles, with Team B selecting a rule. The team which requires the fewest number of guesses is the winner. Guess the rule for the numbers in this example.

x	y
2	7
3	12
6	39
5	28
4	19

9

Motions in Geometry

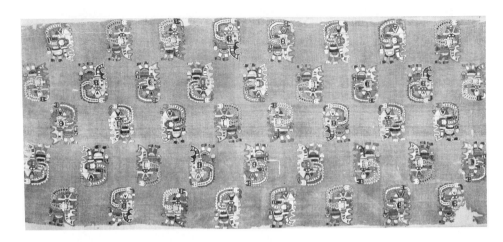

"Mantle," by Paracas Peru (92″ by 35″)
Courtesy Museum of Fine Arts, Boston, Ross Collection.

ACTIVITY SET 9.1

TRANSLATIONS, ROTATIONS, AND REFLECTIONS

Translations, rotations, and reflections are mappings in geometry which can be described by motions. A *translation* is illustrated by a sliding motion. Moving glaciers, flowing rivers, and the shearing of the earth's crust along a fault line are examples of translations. A *rotation* is a turning motion. This occurs in the movements of whirlpools, cyclones, and planets. A *reflection* can be described as a flipping motion or by a mirror image. A landscape and its reflection in a lake or pool produce two identical scenes. It looks as if one scene could be "flipped about" the shoreline to coincide with the other.

Translations, rotations, and reflections for figures in a plane are illustrated by the diagrams shown next. If Figure A is traced on paper, it can be slid to the right until it coincides with Figure B. This is a translation which maps Figure A to Figure B. If Figure C is traced and the paper is turned about point *O,* it can be made to coincide with Figure D. This is a rotation about point *O* which maps Figure C to Figure D. Figure E can be flipped onto Figure F so that they coincide by folding this page about line *M*. This is a reflection about line *M* which maps Figure E to Figure F.

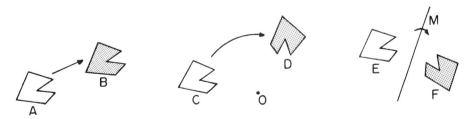

Patterns for wallpaper and fabric can be formed by carrying out combinations of translations, rotations, and reflections on a basic figure. The pattern of little figures in the preceding photograph of the fabric can be obtained by mapping these figures down the diagonals from upper left to lower right. Beginning with the figure in the upper left corner, a horizontal reflection followed by a translation produces the position of the second figure in the diagonal. If the second figure is rotated 180° and then translated, we get the location of the third figure in the diagonal. In this manner each diagonal can be generated by mappings of the figures from the top row of the fabric.

1. **Rotations and Reflections with Toys:**
   The Three-cornered Toy is one of several "toys" described in *Let's Play Math* by Michael Holt and Zoltan Dienes.* One objective of these toys is to provide an informal introduction to rotations and reflections.

   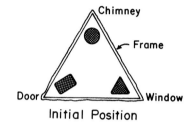
   Initial Position

   Cut out a triangle and label both the front and back corners with a rectangle, triangle, and circle, as shown here. Place the triangle on a piece of paper and draw a frame around it. At each corner of the frame write "chimney," "door," and "window." Each position that this triangle can be placed in the frame, indicates a different arrangement of shapes to be used for the door, chimney, and window of a house. The arrangement of these shapes in the house shown below is from the initial position of the triangle in the frame.

   ★ a. Find the remaining five different positions of the triangle and sketch the corresponding houses.

   ★ b. There are three clockwise rotations (rotations of 120°, 240°, and 360°) and three reflections (reflections about L_1, L_2, and L_3) which map the triangle from the initial position to its six different positions. For example, a reflection about L_1 leaves the "circle" in the top position for the chimney and interchanges the two lower figures so that the "rectangle" is the window and the "triangle" is the door. Under each house write the rotation or reflection which leads to its design.

   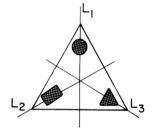

*M. Holt and Z. Dienes, *Let's Play Math* (New York: Walker and Company, 1973), pp. 88-94.

c. Starting from the Initial Position, if the triangle is rotated 120° clockwise and then reflected about line L_1 (see figure in part **b**) it will have the position shown in Figure 1. Will the triangle be in the same position if it is first reflected about L_1 and then rotated 120° clockwise? Try this and sketch the locations of the door, chimney, and window in Figure 2.

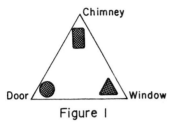

Figure 1

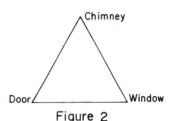

Figure 2

★ d. These types of "toys" can be made with many different shapes and have a variety of uses. This Five-cornered Toy (pentagon) can be used for determining color combinations for dressing. In how many different ways can this pentagon be placed in its frame?

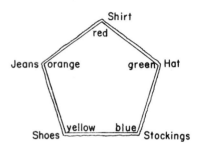

2. **Reflections by Paper Folding:** Fold a piece of paper in half three times. Each time fold perpendicular to the previous fold. Position the paper so that the original four corners are in the top right-hand corner. Draw and cut out a backward L-shaped hexagon.
When the paper is unfolded there will be eight L-shaped hexagons. Try to predict the results by sketching their positions in the eight boxes shown next. Then unfold your paper and check the answers.

★ a. What single mapping will move the hexagon from the lower right position to the lower left?

b. What single mapping will move the hexagon from the lower left to the upper right position?

★ c. What two mappings will move the hexagon from the lower right to box 7?

d. What single mapping will move the hexagon from box 8 to box 6?

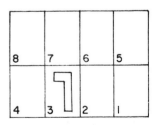

3. **Composition of Mappings:** Any number of mappings can be carried out in succession. For example: a rotation, followed by a reflection, followed by a translation, and so forth. When one mapping is followed by another it is called a *composition* or *product of mappings*. Each mapping moves the image of the previous mapping. For a mapping or composition of mappings, a figure and its image are always *congruent*. By tracing and paper folding determine the final image of the shaded figure below for the given sequence of mappings.

First Mapping	Rotation about O which maps A to A'
Second Mapping	Reflection about line L
Third Mapping	Translation which maps B to B'
Fourth Mapping	Reflection about line M

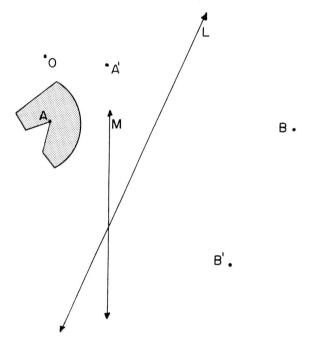

Rotations and Reflections on Geoboards:

4. The circular geoboard (Material Card 18) is more convenient for illustrating rotations than rectangular geoboards or grids. Since there are 24 pegs on the outer "circle," rotations of 15°, 30°, 45°, etc., can be easily represented. Trapezoid T, which is shown on this geoboard, will coincide with trapezoid T' by a clockwise rotation of 120° about the center of the board. Explain why it is easy to find this angle of rotation on a geoboard.

Sketch the image of each set of points for the clockwise rotations given below these geoboards.

a. 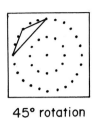 b. 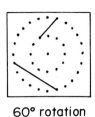 ★ c.

    45° rotation   60° rotation   30° rotation

5. Sketch the images of the figures on the geoboards for reflections about the given line segments.

a. 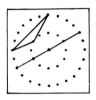 ★ b. c.

6. Triangle T is mapped to triangle T' by a reflection about line L. A second reflection about line M maps triangle T' onto triangle T''.

★   a. What rotation will map triangle T to triangle T''?

     b. Compare the number of degrees in angle XOY with the degrees in the rotation that maps triangle T to triangle T''. What do you find?

     c. Find the image of quadrilateral H by reflecting it about line R and then reflecting its image about line S.

★   d. The result of one reflection followed by another is called a *composition of reflections*. Find a rotation that will have the same effect on quadrilateral H as the composition of the two reflections in part **c**.

     e. The composition of reflections about two intersecting lines is always a rotation about the point of intersection. Use the preceding examples to form a conjecture about how the number of degrees in the rotation is related to the angle formed by the intersecting lines.

ACTIVITY SET 9.1

ACTIVITY SET 9.2

DEVICES FOR INDIRECT MEASUREMENT

It was in the second century B.C. that the Greek astronomer Hipparchus applied a simple theorem from geometry to measure distances indirectly. He computed the radius of the earth, the distance to the moon, and other astronomical distances.* The theorem that Hipparchus applied states that if two triangles are *similar* (have the same shape), the ratio of the lengths of any two sides of one triangle is equal to the corresponding ratio for the other triangle. This important theorem is used in surveying, map-making, and navigation. In the following activities you will see this theorem applied in methods and devices for indirect measurement.

1. **Sighting Method:** Artists hold out their thumbs to estimate the sizes of objects they are painting. Let's examine the theory behind this method. Stretch your arm out in front of you with a ruler held in a vertical position. Select some distant object such as a person and estimate his or her height. Hold the ruler so that the top is in line with your eye and the top of the person's head. Position your thumb on the ruler so that it is in line with the person's feet.

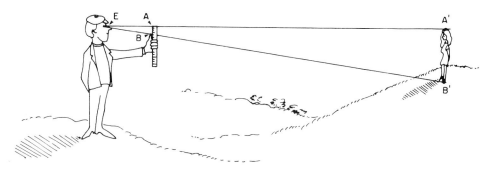

★   a. The lines of sight from your eye (point E) to the top of your ruler and your thumb form triangle EAB. The lines of sight from E to the head and feet of the observed person form triangle $EA'B'$. Why are these triangles similar?

b. Obtain the length of $\overline{EA}$ by measuring the distance from your eye to a ruler as shown in the figure.

c. Since triangle EAB is similar to triangle $EA'B'$, the ratios of the lengths of their corresponding sides are equal. Assume that AB is 4 centimeters and the height of the person is 165 centimeters. Use your measurement from part **b** for EA and the ratios given here to find EA'.

$$\frac{EA'}{EA} = \frac{A'B'}{AB}$$

---

*Several measurements by Hipparchus are described by M. Kline, *Mathematics in Western Culture* (New York: Oxford University Press, 1953), pp. 67–73.

2. **Scout Sighting Method:** Here is an old Boy Scout trick for getting a rough estimate
of the distance to an object. This is a variation of the sighting method but does not
require a ruler. The scout stretches out his arm and, with his left eye closed, sights
with his right eye along the tip of a finger to a distant object. Then he closes his right
eye and by sighting with his left eye sees his finger in line with another object. He then
estimates the distance between the two objects and multiplies by 10 to get his distance
from these objects.*

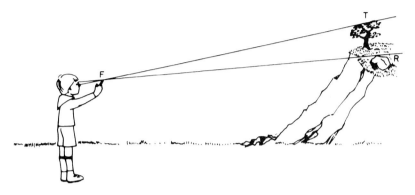

a. If the Boy Scout estimates the distance between the tree (T) and the rock (R) to
be 90 meters (about the length of a football field), how far is he from the tree?

b. In the following diagram E_1 and E_2 represent the eyes, F the finger, T the tree, and
R the rock. The two isosceles triangles that are formed are similar. What is the ratio,
E_1F/E_1E_2, for the measurements on this diagram?

★ c. Explain why 10×90 gives the distance from F to T in meters.

d. For many people the distance E_1F
(from eye to fingertip) is 10 times
E_1E_2 (the distance between the eyes).
Stretch your arm out in front of you
and measure the distance from your
eye to your raised index finger. Then
measure the width between your eyes.

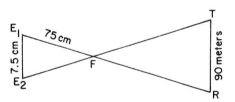

Should you be using the number 10 for the scout sighting method? If not, what number
should you be using?

★ e. Try this method to measure your distance from an object. If you are fairly close
to an object, such as something you might see in a room, you should add the distance from eye to finger to the result of your computation. To approximate
greater distances, the distance from eye to finger, E_1F, can be ignored. Explain why.

*See C.N. Shuster and F.L. Bedford, *Field Work in Mathematics* (East Palistine, Ohio: Yoder Instruments, 1935), pp. 56–57, for several more elementary sighting methods.

3. **Stadiascope:** The stadiascope works on the same principle as the sighting method with the ruler (see Activity 1). This instrument has a peep sight at one end and cross threads at the other. The stadiascope shown here was made from a paper-towel tube having a length of 27 centimeters. At one end of the tube six threads have been taped in horizontal rows. These threads are 1 centimeter apart. A piece of paper has been taped over the other end and a peep sight has been punched on the same level as the lowest thread.

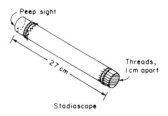

The following diagram shows a stadiascope being used to measure the distance to a truck. The truck fills four spaces at the end of the stadiascope. Let's assume that the truck has a height of 3 meters.

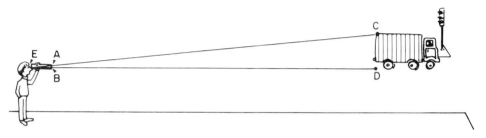

a. Point E is the peep sight, and points A and B mark the upper edge and lower edge of the four spaces of the stadiascope in which the truck is seen. Triangle EBA and triangle EDC are right triangles. Explain why they are similar.

★ b. Since triangle EBA is similar to triangle EDC the ratios of their corresponding sides are equal. Use the ratios given here to compute the length of $\overline{ED}$ in meters. (*Note:* EB and AB can be left in centimeters and CD in meters.)

$$\frac{ED}{EB} = \frac{CD}{AB}$$

★ c. Suppose you sight a person who fills two spaces of the stadiascope and you estimate this person's height to be 180 centimeters. How many meters is this person from you?

d. Make a stadiascope and use it to measure some heights or distances indirectly. Check your results by direct measurements.

4. **Clinometer:** The clinometer is used to measure heights of objects and is an easy device to construct (see Material Card 21). It is a simplified version of the quadrant, an important instrument in the Middle Ages, and the sextant, an instrument for locating the positions of ships. Each of these devices has arcs which are graduated in degrees for measuring angles of elevation. The arc of the clinometer is marked from 0 to 90 degrees. When an object is sighted through the straw, the number of degrees in angle BVW can be read from the arc. Angle BAC is the angle of elevation of the clinometer. What will happen to angle BVW as angle BAC increases?

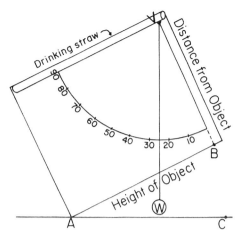

Angle BVW on the clinometer is equal to the angle of elevation of the clinometer, angle BAC. In the following figure the clinometer was used to find the angle of elevation from eye level to the top of the tree. This angle is $18°$. The distance from the person to the base of the tree, ET, is 60 meters.

★ a. Use a ruler to set up a scale on $\overline{ET}$. Determine the distance each unit of your scale represents. Use your scale to find the length of $\overline{TR}$.

b. The observer's eye is 150 centimeters above the ground. What is the height of the tree?

c. Make a clinometer by pasting Material Card 21 onto a solid backing. Use this device to measure the heights of some objects. (*Note:* Draw a similar triangle and set up a scale.)

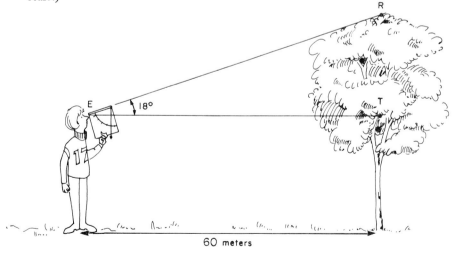

★ d. Sometimes the base of an object is inaccessible. In the picture shown here, point *K* is directly below point *C* and somewhere inside these buildings. Therefore, the horizontal distance from *A* to *K* cannot be measured directly. However, the angles of elevation at *A* and *B* can be used to construct triangle *ABC*. Given that the distance between *A* and *B* is 10 meters, set up a scale on $\overline{AB}$ and determine the vertical height from *K* to *C*.

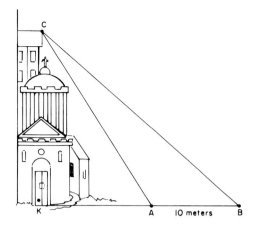

5. **Hypsometer**: The hypsometer is used for the same purpose as the clinometer but it eliminates the need to measure angles (see Material Card 21). The grid is used to set up a scale on the sides of the hypsometer. For example, suppose that you sight to the top of a building and that the string of the hypsometer falls in the position shown here. If you are 55 meters from the base of the building, this distance is represented on the right side of the hypsometer by letting each space be 10 meters. Beginning at distance 55 on the right side of the hypsometer and following the arrow in

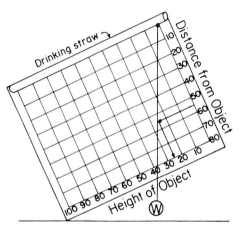

to the plumb line (weighted string) and then down to the lower edge of the grid, shows that the height of the building is 25 meters.

★ a. For the position of the string which is shown on this hypsometer and a distance of 70 meters from an object, what is the height of the object?

b. If you know the height of an object, the hypsometer can be used to find the distance to its base. For the position of the string on this hypsometer, what is the distance to the base of a tree that is 30 meters tall?

c. Make a hypsometer from Material Card 21 and use it to measure the heights of some objects.

6. **Transit:** The transit (or theodolite) is the most important instrument for measuring horizontal and vertical angles in civil engineering. In its earliest form this instrument was capable of measuring only horizontal angles, much like the simplified version shown here. To make this transit, tape a protractor (Material Card 41) to one end of a meter stick (or long thin board) and pin a drinking straw at the center point of the protractor.

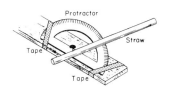

a. The following diagram shows how the transit can be used to find the distance between two points. Line segment $\overline{AB}$ has been marked by posts on the left bank of the river. By holding the stick of the transit parallel to $\overline{AB}$, angle FAB and angle FBA can be measured. The distance between the posts can be found by a tape measure or by pacing off. This distance is 500 meters. Use a ruler to set up a scale on $\overline{AB}$. The distance from A to the fort can be found by using your scale.

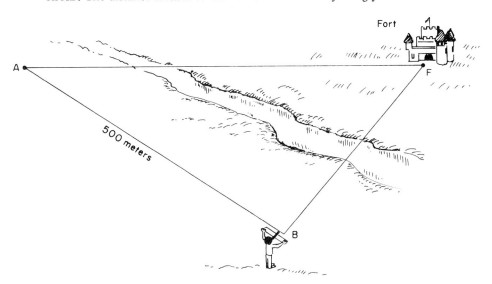

★ b. Explain how you would use the transit to make a map of the objects in this field. What is the least number of angles and distances you would have to measure?

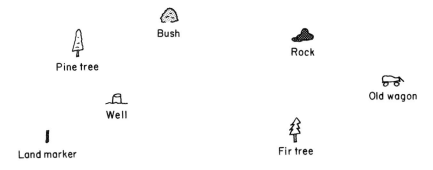

ACTIVITY SET 9.2 179

7. **Plane Table**: The Plane Table is one of the principal instruments used by the United States Coast and Geodetic Survey for mapping. On top of this table there is a straightedge, called an *alidade*, which has sighting posts at each end. This remarkably simple device can be used to make a map without measuring angles.

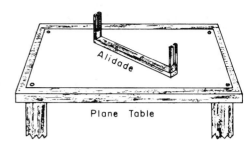
Plane Table

To map the location of objects, set the table at a convenient location and tape a sheet of drawing paper to its surface. Select a point on the paper as a projection point and label it *O*. Then line the alidade up with point *O* and the most distant object to be mapped, and draw a line from *O* toward this object. Measure the actual distance to the object and set up a scale on the drawing paper to plot a point for this object. Draw lines from *O* to the other points to be mapped. Measure their distances and use the same scale to plot these points on the paper.

   a. Here are the same field objects that appear in Activity 6. Using your ruler as an alidade and point *O* as projection point, make a map of the location of these objects inside the given rectangle.

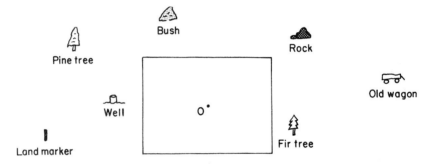

   b. Tape a piece of paper to a table and use the plane table method to make a map giving the locations of several objects in a room. The location of the projection point *O* is arbitrary. The edge of a ruler can be used as an alidade.

Eight-foot steel Moebius band by Rivera—rotates once every six minutes—
History and Technology Building, Smithsonian Institute
Smithsonian Institute Photo no. 72-233.

ACTIVITY SET 9.3

TOPOLOGICAL ENTERTAINMENT

1. One of the many stories about the legendary Paul Bunyan involves the Moebius band. Paul was using a Moebius-band-shaped conveyor belt to bring out uranium ore from a mine in Colorado. The belt ran 1/2 kilometer into the mine and was a little over 1 meter wide. After several months of operation the mine gallery had become twice as long and Paul decided to cut the belt down the middle to increase its length. He told Ford Fordsen to take his chain saw and cut the belt lengthwise.

"That will give us two belts," said Ford Fordsen. "We'll have to cut them in two crosswise and splice them together. That means I'll have to go to town and buy the materials for two splices." "No," said Paul. "This belt has a half twist—which makes it what is known in geometry as a Moebius strip." *

★ a. Make a Moebius band from a strip of paper by putting a half-twist in it and taping together the two ends of the strip. Cut it in two lengthwise as pictured here. Who was right, Fordsen or Bunyan?

★ b. The Moebius band has only one side. Is the band you get in part **a**, after cutting, one or two sided? (To check, draw a line lengthwise around the band until it returns to the starting point.)

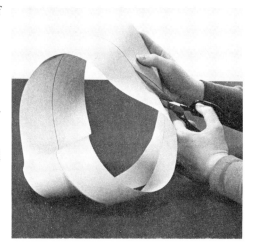

c. Later in the story Paul cut the conveyor belt lengthwise a second time to make it even longer. This time a character called Loud Mouth Johnson disagreed with Paul on the number of pieces which would be produced. Use your band from part **b** and cut it lengthwise. Will you get one or two bands? Check the number of sides.

2. The B.F. Goodrich Company manufactures the Turnover Conveyor Belt System. This belt has two half-twists and has an advantage over conventional belt systems in that only the clean side of the belt is in contact with the idlers on the underside.

*W.H. Upson, "Paul Bunyan versus the Conveyor Belt, *Ford Times,* Ford Motor Company.

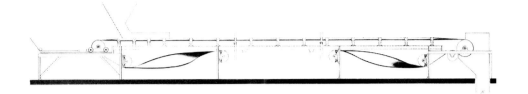

   a. Make a band from a strip of paper by putting in two half-twists and taping together the ends of the band. How many sides does it have?
   b. Form a band with three half-twists. How many sides does it have?
   c. Make a conjecture about the number of sides of bands with even and odd numbers of twists.

3. If you were surprised by the results in Activity 1 you may wish to try predicting the outcomes for the following types of cuts. After each cut describe the number of bands and the number of sides to each band.

★ a. Form a Moebius band. Draw a line by beginning one-third of the way in from one edge and continuing until you return to the starting point. This one continuous line will produce "two tracks" as shown in this drawing. What will be produced by cutting along this line?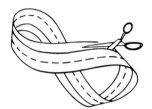

   b. Use the band with two half-twists which you made in part **a** of Activity 2. What will happen when this band is cut lengthwise?
   c. Use your band from part **b** of Activity 2. Try to predict what you will get when it is cut lengthwise.

4. **Double Moebius Band:** Place two strips of paper together (as pictured here), give them a half-twist (as if they were one strip), and tape the edges together. The result is a double Moebius band. Run your fingers or a pencil all the way around the double band between the two layers. Is this two separate bands or one band?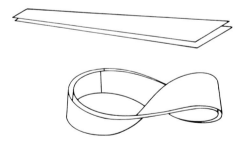

   a. What will happen when the double band is opened? Describe the results and the number of sides.
   b. Make a triple Moebius band out of three strips of paper. Try to predict what will happen when it is opened. Describe what you get. (*Note:* When the 6 ends are brought together, they are taped in pairs: top 2, middle 2, and bottom 2.)

★ 5. **Gale Game:*** This game was devised by David C. Gale of Brown University and is played on a grid of alternately colored rows of dots (in this case white and black). One player connects only white dots and the other player connects only black dots. The players take turns connecting any two adjacent dots. The objective is to construct a path from one edge of the square to the other, with one player trying to connect the top edge of the grid to the bottom and the other player trying to connect the left edge to the right.

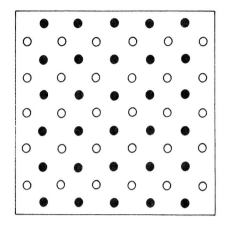

Only one line between two dots is drawn on each turn and no line is permitted to cross an opponent's line. Try playing this game. Is it possible for two players to block each other so that no one wins, or will there always be a winner?

★ 6. **Party Trick:** Cut two pieces of string with lengths of about 50 centimeters. Using one of these strings, tie each end around someone's wrists. Then tie another person's wrists with the second piece of string so that they are linked together as shown in this picture. The object is to get unlinked without untying or cutting the string. Try it. Explain how this can be done.

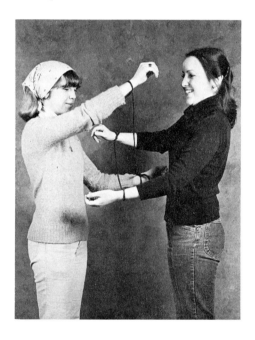

*The Gale Game and the Card and Ring Puzzle (Activity 8) are described by M. Gardner, "Four Mathematical Diversions Involving Concepts of Topology," *Scientific American,* **199** No. 4 (April 1958), pp. 124–29.

10

Probability and Statistics

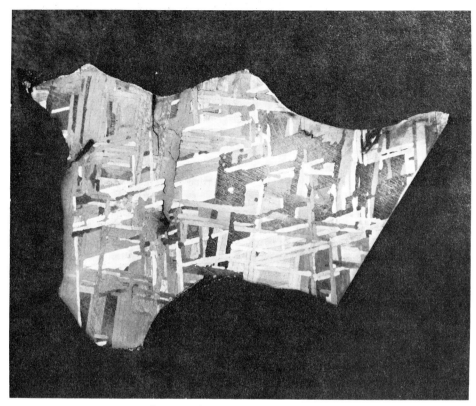

Cross section of a stone meteorite

"He who has heard the same thing told by 12000 eye-witnesses has only 12000 probabilities, which are equal to one strong probability, which is far from certainty."

Voltaire

ACTIVITY SET 10.1

PROBABILITY EXPERIMENTS

Graphical analysis is a procedure for plotting the positions of storms, aircraft, atmospheric winds, meteors, meteorites, and other objects by using many sources of information. For example, when a large meteorite is sighted there are dozens of phone calls and inquiries to news services. Most of the reports of distance, direction, and angular altitude contain erroneous information. However, by "averaging," graphing, and analyzing the data, scientists obtain the most probable direction of the meteorite and point of impact. While the conclusions from this technique are far from certain, they are, in the words of Voltaire, "equal to one strong probability."

In regard to a great many events, there is in the mind of everyone a state of uncertainty. Probability is the branch of mathematics which deals with determining the degree of belief which should be assigned to these events. There are two types of probabilities: *theoretical probability,* which is determined solely by mathematical calculations; and *empirical probability,* which is determined by observing the outcomes of experiments. Let's consider some examples. The theoretical probability of rolling a 4 on a die is 1/6, because there are six faces on a die, each with an equally likely chance of turning up, and only one of these faces has a 4. An empirical probability for obtaining a 4 can also be found by rolling a die many times and recording the results. If we divide the number of times a 4 is rolled by the number of rolls, the quotient is an empirical probability. Of the thousands of meteorites which have been found, about 9 out of 10 are made of rock and the others are made of iron. Based on this observation, the empirical probability that the next meteorite which hits the earth will be rock is 9/10.

1. Suppose someone wanted to bet that a tack will land with its point up when dropped on a hard surface. Would you accept the bet? Before you wager any money try the following experiment. Take ten identical tacks and drop them on a hard surface. Count the number of tacks landing point up. Do this experiment 10 times and record your results in the table.

Trial number	1	2	3	4	5	6	7	8	9	10
Number with points up										

   a. Out of the 100 tacks you dropped what was the total number of tacks landing point up? On the basis of this experiment explain why you would or would not accept the bet.

   b. What is the empirical probability of this type of tack landing point up?

   $$\frac{\text{Number with point up}}{\text{Total number dropped}} =$$

   c. About how many tacks would you expect to land point up if you had dropped 300?

   d. What is the empirical probability that this type of tack will land with its point down? How is this probability related to the probability in part **b**?

2. Games of chance like the following one are common at carnivals and fairs. A penny is spun on a square grid, and if it lands inside a square you win a prize. Use the Two-Penny Grid (Material Card 12) to try this game. Compare the diameter of a penny with the width of a square. Do you think there is a better chance that the penny will land inside a square or on the lines?

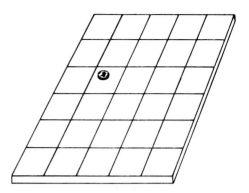

a. Spin a penny 20 times and tally the results.

Penny inside square	Penny on line

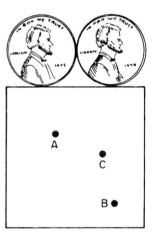

b. Based on your experiment, what is the empirical probability that a penny will land inside a square?
c. The length of the sides of the squares is twice the diameter of a penny. Which of the points that are labelled A, B, and C will be winning points if they represent the center of a penny?
★ d. Shade in the region of this square in which the center of every winning penny must fall. What fractional part of the square is this shaded region?
★ e. Compare the shaded region to the whole square to determine the theoretical probability that a penny will land inside a square. Is this a fair game, in the sense that there is a 50–50 chance or better of winning?

3. The sides of the squares on the Three-Penny Grid (Material Card 16) are 3 times the diameter of a penny. Use this grid to try the spinning game. Do you think a carnival would win or lose money on this grid?

   a. Spin a penny 20 times and compute the empirical probability of a winning penny.

Penny inside square	Penny on line

   ★ b. Shade in the region of the square in which the centers of winning pennies must fall. What is the theoretical probability of a winning penny?

   ★ c. What is the probability that a spinning penny will not touch the lines on a Four-Penny Grid?

4. Make three cards of the same size and label both sides of one with the letter A, both sides of the second with the letter B, and one side of the third with A and the other side with B. Select a card at random and place  it on the table. The card on the table will have either an A or a B facing up. What is the probability that the letter facing down will be different from the letter facing up? Take an intuitive guess. Repeat this activity 24 times and record the number of times that the letter facing down is different from the letter facing up. Then compute the empirical probability.

Same letter facing down	Different letter facing down

   Once a card has been selected and it is clear which letter is facing up, people often feel that there is a 50–50 chance that the letter facing down will be different. This is a false assumption. The questions that follow in parts **a** through **e** will help you to determine the correct probability.

ACTIVITY SET 10.1   191

There are six possible outcomes when a card is selected. To distinquish between the two sides of the A,A card, let's refer to one side as A_1 and the other as A_2. Similarly, the two sides of the B,B card will be referred to as B_1

Outcome	1	2	3	4	5	6
Top face	A_1	A_2	B_1	B_2	A	B
Bottom face	A_2	A_1	B_2	B_1	B	A

and B_2. (*Note:* We will not actually change the labels on these cards.) The six possible outcomes for the top and bottom faces are listed in the above table.

★ a. In how many of these outcomes is the top face of the card different from the bottom face? (Remember, the faces are labelled with only A's and B's.)

b. What is the probability of selecting a card with the letter on the top face different from the letter on the bottom face?

★ c. Suppose that a card has been selected and an A is face up. List the three outcomes from the table for this event.

d. What is the probability that the letter on the bottom face of the card in part c will be a B?

★ e. Is it any help in determining the probability that the two letters will be different, if you know the letter facing up on the selected card?

Just for Fun—Buffon's Needle Problem

In 1733 the French naturalist Count Buffon presented the following problem to the Academy of Sciences in Paris.

If a needle of length L is tossed onto a surface of parallel lines, where the distance between the lines is D, what is the probability that the needle will come to rest on a line?

In the following experiment we will use a needle whose length (L) equals the distance (D) between the lines. Look at the above figure and try to predict the probability of a needle coming to rest on a line. Is this probability greater than or less than 1/2?

Cut a stick (toothpick, match, etc.) of length 1 centimeter and drop it onto the lined surface on Material Card 19. (The distance between the parallel lines is 1 centimeter.) Repeat this activity 100 times and count the number of times the stick rests on a line.

a. Compute the empirical probability of this event.

b. The theoretical probability of Buffon's needle problem was first computed in 1777 and, surprisingly, it involved the number π.

$$\text{Probability} = \frac{2L}{\pi D}$$

In your experiment $L = 1$ and $D = 1$, so the theoretical probability is $2/\pi$. Compute $2/\pi$ to two decimal places and compare it to your empirical probability (use $\pi = 3.14$).

★ c. For sufficiently large numbers of tosses this experiment can be used to approximate π. In 1901 the Italian mathematician Lazzerini had 2169 "hits" in 3408 tosses. What was his empirical probability for this experiment? Use this probability in the following equation to determine the value of π (probability $= 2/\pi$).

d. Use your probability from part **a** and the equation in part **c** to determine a value for π.

★ e. The empirical probability in part **a** will often be close to $2/\pi$, the theoretical probability. What will be the theoretical probability if the distance between the lines is doubled to 2 centimeters? Repeat the previous experiment by using every other line on the material card. Compare your empirical probability with the theoretical probability.

ACTIVITY SET 10.2

COMPOUND PROBABILITY EXPERIMENTS

The probability of two or more events occurring is called a *compound probability*. When one event does not influence or affect the outcome of another, the two are called *independent events*. Otherwise, they are *dependent events*. Consider the probability of drawing 2 red chips from a box containing 3 red chips and 3 white chips. On the first draw, the probability of obtaining a red chip is 1/2. Now, if the first chip is replaced before the second draw, the probability of drawing another red chip will be 1/2, since the number of red chips is the same as it was on the first draw. These two drawings are independent events because the result of the first draw does not affect the outcome of the second draw. The probability of drawing the 2 red chips is the product of the two probabilities 1/2 × 1/2, or 1/4. A completely different situation exists if we do not return the chip to the box after the first draw before drawing the second chip. In this case, the probability of drawing a red chip on the second draw depends on what happens on the first draw. That is, these events are dependent. For example, if a red chip is taken on the first draw, the probability of selecting a red chip on the second draw will be 2/5. In this case, the probability of drawing 2 red chips is 1/2 × 2/5, or 1/5. In the following activities there are several experiments involving compound probabilities with both independent and dependent events. In each experiment the empirical probability will be compared with the theoretical probability.

1. The following experiment has a result which, most likely, will be contrary to your intuition. Place 3 red and 3 white chips in a box and select 2 of them at random (both at once). It seems reasonable to expect a 50% chance of getting 2 chips of different colors. Repeat this experiment 30 times and tally your results.

Two chips of different color	Two chips of same color

   a. Based on your results, what is the empirical probability of drawing 2 chips of different color?

   ★ b. Drawing 2 chips is the same as drawing 1 and then (without replacing it) drawing a second. These are dependent events. What is the probability of drawing a red chip and then a white? (*Hint:* Multiply the probabilities of these two events.)

   c. What is the probability of drawing a white chip first and then (without replacing it) a red?

   ★ d. The theoretical probability of drawing 2 chips of different colors (regardless of order) is the sum of the probabilities in parts **b** and **c**. What is this probability?

   e. Another method of determining the theoretical probability in part **d** is to list the 15 possible pairs of chips, as shown below. How many of these pairs have a red and a white chip? Use this number to determine the probability of selecting two chips of a different color. Compare this answer to your empirical probability from part **a**.

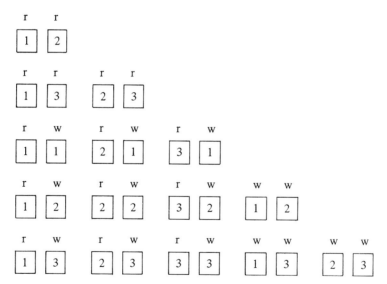

194   CHAP. 10 PROBABILITY AND STATISTICS

2. What is the chance that on a toss of 4 coins, 2 will be heads? If you are one of the people who feel that it is 1/2, you may be surprised at the results in the following experiment. Toss 4 coins and record the number of heads and tails. Do this 32 times.

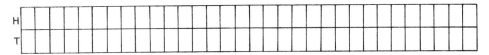

a. According to your table, what is the empirical probability of exactly 2 heads in a toss of 4 coins?

★ b. Tossing 4 coins together is like tossing a single coin 4 times. Continue this tree diagram to show all possible outcomes when a coin is tossed 4 times in succession. There will be 16 possible outcomes. How many of these will have 2 heads and 2 tails?

★ c. What is the theoretical probability of getting 2 heads and 2 tails in a toss of 4 coins?

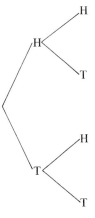

First Toss   Second Toss

★ 3. The chances of succeeding in compound events are often deceiving. You might expect that if a box has 7 red chips and 3 white chips, your chance is fairly good of drawing 2 red. Try this experiment 30 times and record your results. Then compute the theoretical probability of drawing 2 red. (*Hint:* Multiply the probability of drawing the first red chip by the probability of drawing the second red chip.)

2 red chips	1 or more white chips

ACTIVITY SET 10.2

★ 4. Were you surprised in Activity 3 that the chance of drawing 2 red chips is so poor? Even by replacing the first chip before drawing the second, the chances of getting 2 red chips are less than you might expect. Carry out the same experiment by drawing the chips separately and replacing the first chip each time. Repeat this procedure 30 times and record the results. Then compute the theoretical probability of getting 2 red chips by replacing the first chip. (*Hint:* Multiply the two probabilities for drawing each chip separately.)

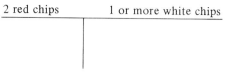

5. Do you think that in 24 rolls of 2 dice at least one double 6 would be likely to occur? This is one of the seventeenth century gambling questions which led to the development of probability. Toss the dice 24 times and record the pairs of numbers on each toss.

Record of 24 Tosses

Did you roll a double 6? Even though this event may have occurred one or more times in your experiment, there is slightly less than a 50% chance of its happening. Use the following steps to determine the theoretical probability of this event. This will be done by first determining the probability of not obtaining a double 6 in 24 rolls. The probability that an event will happen and the probability that it will not happen are called *complementary probabilities*.

★ a. What is the probability of rolling a double 6 on one toss of the dice?

b. What is the probability of not rolling a double 6 on one toss of the dice?

★ c. What is the probability of not rolling a double 6 on two rolls of the dice?

d. Indicate how you would find the probability of not rolling a double 6 in 24 rolls of the dice. (Do not multiply your answer out.)

★ e. The product $35/36 \times 35/36 \times \cdots \times 35/36$, in which $35/36$ occurs 24 times, is approximately .51. What is the probability of rolling at least one double 6 in 24 rolls of the dice?

Just for Fun—Trick Dice

The 4 dice shown next have the following remarkable property: No matter which die your opponent selects, it is always possible to select one of the remaining 3 dice, so that the probability of winning is in your favor. Make a set of these dice, either by "doctoring up" regular dice or using the dice on Material Card 42. Try this experiment with Die 2 and Die 4. Can you tell which of these two dice is more likely to win? Roll each die 21 times and record the numbers in the appropriate boxes of the following table. Circle the numbers for which Die 4 wins over Die 2.

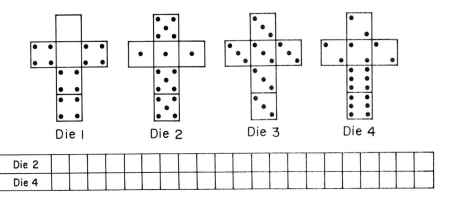

Die 2																																				
Die 4																																				

a. Compute the empirical probability for the winning die.

★ b. Complete this table to show all 36 pairs of numbers from the 2 dice. Circle the pairs for which Die 4 has the greater number. What is the theoretical probability that Die 4 wins over Die 2?

★ c. If a 2 is rolled on Die 4 and a 1 is rolled on Die 2, then Die 4 wins. What is the probability of these events happening? (*Hint:* Multiply the two probabilities.)

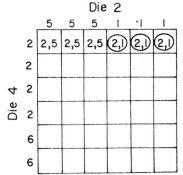

★ d. If a 6 is rolled on Die 4, it wins over every number on Die 2. What is the chance of a 6 on Die 4?

★ e. The probability of Die 4 winning over Die 2 is the sum of the probabilities in parts **c** and **d**. What is this probability?

Similar approaches can be used to show that: Die 2 wins over Die 1; Die 1 wins over Die 3; and Die 3 wins over Die 4.

© 1965 United Feature Syndicate, Inc.

ACTIVITY SET 10.3

APPLICATIONS OF STATISTICS

The word "average" has several meanings. Loosely speaking, it means *typical* or *usual* and is used in this way in the above cartoon. In mathematics there are three averages for describing a set of numbers: mean, median, and mode. Let's use these words to describe the following numbers: 2, 3, 4, 6, 8, 11, 12, 12, 12, and 18. The *mean* is what is commonly referred to as "the average." It is the sum of the values divided by the total number of values. The sum of the ten numbers is 88, so the mean is 8.8 (88 ÷ 10). The *median* is the middle number when the numbers are arranged in order. The median for these numbers is 9.5, the number that is halfway between the two middle numbers. The *mode* is the value which occurs most frequently, if there is such a value. For the given numbers the mode is 12. Sometimes there is more than one mode. The range and standard deviation are two measures for describing the amount of spread or dispersion in a set of numbers. The *range* is the difference between the largest and the smallest numbers in the set. The range for the preceding numbers is 18 − 2, or 16. The *standard deviation* measures the spread of values from the mean. The steps for computing the standard deviation are listed in Activity 2.

1. Robert Wadlow, who was born in Alton, Illinois, on February 22, 1918, reached a height of 268 centimeters at age 18. His shoes were size 37AA and his hand measured 32 centimeters from the wrist to the tip of his middle finger. How does your hand measurement from wrist to tip of middle finger compare to Robert Wadlow's? Is it half as big?

    a. Collect the hand measurements from the wrist to the tip of the middle finger of your classmates. Record these measurements to the nearest centimeter. (The frequency is the number of students with each measurement.)

    b. Compute the mean, median, and mode of these measurements.

    c. What would be the mean, median, and mode if Robert Wadlow were a member of your class?

Measurement	Frequency
22 cm	
21 cm	
20 cm	
19 cm	
18 cm	
17 cm	
16 cm	
15 cm	

★ 2. Robert Wadlow's handspan (maximum distance between tip of thumb and tip of little finger) was 42 centimeters. What is your handspan to the nearest centimeter? About how many times greater was Wadlow's handspan than yours?

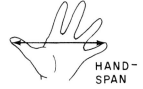
HAND-SPAN

The next table contains the handspans to the nearest centimeter of 24 college students. The mean of these spans is 20.5. Use the following steps for computing the standard deviation. Steps 1 and 2 can be done by completing the table.

Table {
1. Compute the differences between the spans and the mean.
2. Square the differences.
}
3. Compute the mean (average) of the squares of the differences, _____.
4. Find the square root of this mean, _____. This number is the standard deviation for these 24 handspans.

Handspans of 24 People (to the nearest centimeter)

Spans	24	23	23	22	22	22	21	21	21	21	21	21	20	20	20	20	20	20	19	19	19	18	18	17
Differences	3.5	2.5	2.5	1.5									.5						1.5					
(Differences)²	12.25	6.25	6.25	2.25									.25						2.25					

★ a. How many of the spans in the table are within plus or minus 1 standard deviation of the mean? What percent are these spans of the total?

★ b. What percent of the spans are within plus or minus 2 standard deviations of the mean?

c. Obtain the handspans to the nearest centimeter of 12 people and complete the accompanying table. Use the previous four steps to compute the standard deviation. (Round your mean off to the nearest tenth of a centimeter.)

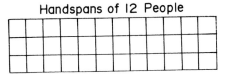
Handspans of 12 People

d. Is your handspan within plus or minus 1 standard deviation of the mean? What percent of these handspans are within this interval?

3. The objective of sampling is to draw conclusions about a large population. The samples are usually relatively small. For example, some national television ratings are based on a sample of only 1100 people. The following activity is designed to show how accurately the sample can represent the given population.

Have a classmate put any one of these three populations of markers in a container, without your knowing the number of red and blue markers.

Population 1	Population 2	Population 3
25 red markers	20 red markers	10 red markers
5 blue markers	10 blue markers	20 blue markers

ACTIVITY SET 10.3

Draw a marker and check off its color in the table. Return the marker to the container and repeat this activity until you have selected 25 markers.

Twenty-five Samples of Markers

Red																									
Blue																									

a. Compute the percentage of red markers for each of the following groups.

First 5 markers   First 10 markers   First 15 markers   First 20 markers   All 25 markers

_____ _____ _____ _____ _____

b. As the size of the sample is increased, there is a greater likelihood that it represents the population. Based on your sample of 25 markers, which of the three given populations do you think you have sampled? Would you have selected the same population if your sample size had been only the first 10 markers?

4. Write the names of the following living areas and their incomes on 18 separate pieces of paper and place them in a container. The experiments that follow will illustrate the need for stratified sampling techniques.*

Rural 9,000	Rural 11,000	Rural 8,000	Urban 11,000	Urban 10,000	Urban 13,000	Urban 9,000	Urban 12,000	Urban 10,000
City 13,000	City 14,000	City 10,000	City 14,000	City 12,000	City 22,000	City 19,000	City 15,000	City 24,000

a. Select six pieces of paper from the container and compute the average (mean) income of this sample. The average of all 18 incomes is $13,111.11. Compare this average with the one you obtained.

★ b. There are many different samples of 6 incomes that you could have selected. Suppose, for example, your sample contained the 6 largest incomes. What is the average of these 6 incomes? What is the average of the 6 smallest incomes?

c. The average of the 6 greatest incomes and the average of the 6 smallest incomes differ significantly from the average of all 18 incomes. By using a technique called *stratified sampling,* you can avoid getting an unrepresentative sample. Separate the urban, rural, and city salaries into three different containers. To obtain a proportional stratified sample of 6 incomes, select 1 rural income, 2 urban incomes, and 3 city incomes. Compute the average of these 6 incomes. Compare this average with the average of the 18 incomes.

★ d. Using the stratified sampling procedure in part c, what is the greatest possible average that can be obtained for 6 incomes? What is the smallest possible average that can be obtained? Compare these averages with those in part b to see the advantage of using stratified sampling.

*See M. Slonim, *Sampling* (New York: Simon and Schuster, 1960), for more examples of stratified sampling.

ACTIVITY SET 10.4

STATISTICAL EXPERIMENTS

A frequency distribution is a tabulation of values by categories or intervals. For example, the births of babies for one year can be listed by: number of births per month; number of births per state; number of births for each age of the mothers; weights and heights of babies; etc. There are several types of frequency distributions which occur commonly. If each category has "approximately" the

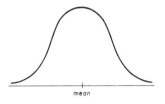

same number of values, the distribution is called a *uniform distribution* (For an example, see Activity 3.a.) The number of births per month for a hospital might be a uniform distribution. One of the most important distributions in statistics is the *normal distribution*. This distribution often occurs in the physical, social, and biological sciences. The graph of a normal distribution is the familiar bell-shaped curve (see figure). This curve is symmetric about the mean with a gradual decrease at both ends. The weights and heights of a large sample of newborn babies will be normally distributed.

★ 1. When 4 coins are tossed there are 6 possible combinations of 2 heads. Think of the coins as being numbered from 1 to 4. Coins 1 and 2 may be heads and coins 3 and 4 tails; or coins 1 and 3 may be heads and coins 2 and 4 tails; etc. Fill in the remaining 4 combinations.

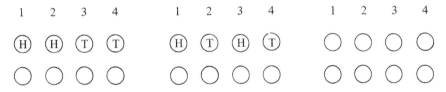

This graph shows the number of different outcomes (combinations of heads and tails) for tosses of 4 coins at a time. The frequency of 6 above 2 heads is for the 6 different combinations of 2 heads. There are frequencies of 1, above 0 heads and 1 head, because these two events can occur only if all coins are tails or all coins are heads. This distribution can be approximated by a normal curve. As the number of coins being tossed is increased, the distributions of outcomes theoretically get closer to a normal curve.

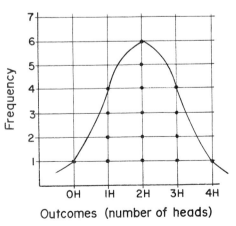

Try the following experiment with 6 coins. Toss the coins and record the number of heads on each toss by placing a dot on the graph. Repeat the tossing of 6 coins 64 times.

a. Form a smooth curve by connecting the top dots on your graph. Is this "close to" a normal curve?

b. The probability of getting 2, 3, or 4 heads is about 78%. What percent of your outcomes had 2, 3, or 4 heads?

c. The probability of getting 5 heads is about 10%. What percent of your outcomes had 5 heads?

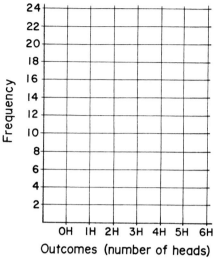

2. Large numbers of measurements from people, such as weight, range of eyesight, IQ, etc., will usually be normally distributed.

    a. This list contains the pulse rates of 60 students. Complete the bar graph on the following page for the intervals on the horizontal axis. Is this "close to" a normal distribution?

 ★ b. This set of pulse rates has a mean of 72 and a standard deviation of about 8.7. What percent of these rates fall within 1 standard deviation above and 1 standard deviation below the mean?

 ★ c. What percent of these rates are within 2 standard deviations of the mean?

Pulse Rates (beats per minute)	Frequency
53	1
56	2
57	1
58	1
61	1
62	3
63	1
64	1
65	3
66	1
67	2
68	2
69	2
70	6
71	1
72	4
73	3
74	4
75	2
76	3
77	1
78	1
79	2
80	3
81	1
82	1
84	2
86	2
89	1
91	1
92	1
Total	60

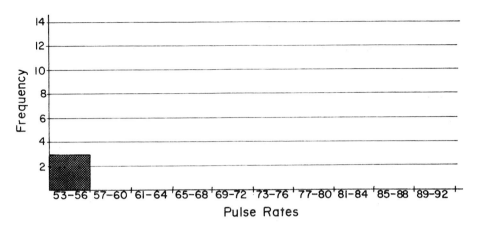

d. Even smaller numbers of measurements may tend to be normally distributed. Draw a bar graph of the pulse rates for 10-second intervals of your classmates. Compute the percentages of the numbers of pulse rates for each of the rates from 8 to 15.

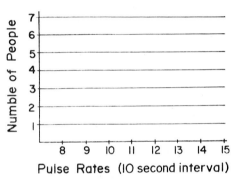

3. Can lists of random digits be produced by writing them down as fast as they pop into your head? Let's see how good you are at doing this. Make a sequence of 100 "random" digits by writing 20 five-digit numbers.

Five-Digit Numbers

1.   6.   11.   16.
2.   7.   12.   17.
3.   8.   13.   18.
4.   9.   14.   19.
5.   10. 15.   20.

a. Find the frequency of each digit in your list and plot it on the graph. If each digit occurs from 7 to 13 times then you have a uniform distribution.* Did you tend to favor or neglect any digits?

b. Describe a method for producing a set of random digits.

c. Use your method in part **b** to obtain 100 digits. Find the frequency of each digit. Did your system favor or neglect any digits?

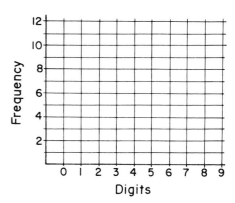

4. Some people are not aware of the uneven distribution of letters which occur in words. One such person may have been Christofer Sholes, the inventor of the typewriter. He gave the left hand 56 percent of all strokes, and the two most agile fingers on the right hand hit two of the least often used letters of the alphabet, j and k. This table shows the frequencies in terms of percentages of the occurrence of our letters in large samples.

E	12.3%	L	4.0%	B	1.6%
T	9.6%	D	3.7%	G	1.6%
A	8.1%	C	3.2%	V	0.9%
O	7.9%	U	3.1%	K	0.5%
N	7.2%	P	2.3%	Q	0.2%
I	7.2%	F	2.3%	X	0.2%
S	6.6%	M	2.2%	J	0.1%
R	6.0%	W	2.0%	Z	0.1%
H	5.1%	Y	1.9%		

★ a. There are 350 letters in the preceding paragraph. Count the number of times that e, s, c, w, and k occur and compute their percentages. Compare your answer with the previous table.

e	s	c	w	k

b. In 1939 a 267-page novel entitled *Gadsby* was released by the Wetzel Publishing Company. The novel was known not for its literary merit but rather for one distinctive feature. Not one of its 50,000 words contained the letter e. Try to compose a sentence of at least ten words which does not use the letter e.

---

*Such limits as these (7 and 13) for determining if a distribution is uniform, can be obtained from a statistical test called the Chi-Square Test.

★ c. In Morse code, letters are represented by dots and dashes so they can be sent by electrical impulses. A dot consumes 1 time unit, a dash 3 time units, and 1 unit is needed for each space between symbols. In 1938 Samuel Morse assigned the symbols with the shortest time intervals to the letters that occurred most frequently, according to his sample of 12,000 words from a Philadelphia newspaper. While Morse's code assignment was more efficient than if he had assigned letters haphazardly,

Code Symbols	Time Units	Letters	Reassigned Code Symbols
·	1	e	
−	3	t	
· −	5	a	
− − −	11	o	
− ·	5	n	
· ·	3	i	
· · ·	5	s	
· − ·	7	r	
· · · ·	7	h	

it still could be improved. Devise a new and more efficient use of Morse's code symbols for the nine letters in this table. How many fewer time units do you require for these nine letters than Samuel Morse did?

5. It is surprising to learn that the leading digits of numbers are not evenly distributed.* To illustrate this fact try the following chance experiment with someone. Find a book that contains a table of numbers (areas or populations of cities, lengths of rivers, etc.). You may wish to use the section of the population table from the 1977 *World Almanac* on the following page. For each number beginning with 1, 2, 3, or 4 you win, and for each number beginning with a 5, 6, 7, 8, or 9 your opponent wins. Even though you are winning on fewer digits you should come out ahead in the long run. The following questions will help you to understand why.

*An explanation of this can be found in W. Weaver, *Lady Luck* (New York: Doubleday, 1963), pp. 270–77.

U.S. Population—Places of 5,000 or More

Place	1970	Place	1970
Haddon Heights	9,365	Palmyra	6,969
Haledon	6,767	Paramus	26,381
Hammonton	11,464	Park Ridge	8,709
Hanover Twp.	10,700	Passaic	55,124
Harrison	11,811	Paterson	144,824
Hasbrouck Heights	13,651	Paulsboro	8,084
Hawthorne	19,173	Pennsauken Twp.	36,394
Hazlet Twp.	22,239	Penns Grove	5,727
Highland Park	14,385	Pennsville	11,014
Hightstown	5,431	Pequannock Twp.	14,350
Hillsdale	11,768	Perth Amboy	38,798
Hillside Twp.	21,636	Phillipsburg	17,849
Hoboken	45,380	Pine Hill	5,132
Hopatcong	9,052	Piscataway Twp.	36,418
Hopewell Twp. (Mercer)	10,030	Pitman	10,257
Irvington	59,743	Plainfield	46,862
Jackson Twp.	18,276	Pleasantville	13,778
Jersey City	260,350	Point Pleasant	15,968
Keansburg	9,720	Pompton Lakes	11,397
Kearny	37,585	Princeton	12,331
Kendall Park	7,412	Princeton North	5,488
Kenilworth	9,165	Prospect Park	5,176
Keyport	7,205	Rahway	29,114
Kinnelon	7,600	Ramblewood	5,556
Lake Hiawatha	11,389	Ramsey	12,571
Lake Mohawk	6,262	Randolph Twp.	13,296
Lake Parsippany	7,488	Raritan	6,691
Lakewood	17,874	Red Bank	12,847
Laurence Harbor	6,715	Ridgefield	11,308
Laconia	8,847	Ridgefield Park	13,990
Lincoln Park	9,034	Ridgewood	27,547
Linden	41,409	Ringwood	10,393
Lindenwold 1973	16,265	River Edge	12,850
Linwood	6,159	Riverside Twp.	8,591
Little Falls Twp.	11,727	Rochell Park Twp.	6,380
Little Ferry	9,064	Rockaway	6,383
Little Silver	6,010	Roselle	22,585
Livingston Twp.	30,127	Roselle Park	14,227
Lodi	25,163	Rumson	7,421
Long Branch	31,774	Runnemede	10,475
Lyndhurst Twp.	22,729	Rutherford	20,802
Madison	16,710	Saddle Brook Twp.	15,975
Magnolia	5,893	Salem	7,648
Mahwah Twp.	10,800	Sayreville	32,508
Manville	13,029	Scotch Plains Twp.	22,279
Maple Shade Twp.	16,464	Secaucus	13,228
Maplewood Twp.	24,932	Somerdale	6,510
Margate City	10,576	Somers Point	7,919
Marlboro Twp.	12,273	Somerville	13,652
Marlton	10,180	South Amboy	9,338
Matawan	9,136	South Orange	16,971
Maywood	11,087	South Plainfield	21,142
McGuire	10,933	South River	15,428
Mercerville-Hamilton Sq.	24,465	Sparta Twp.	10,819
Metuchen	16,031	Spotswood	7,891
Middlesex	15,038	Springfield Twp.	15,740

★ a. The populations on this page are all four-, five-, or six-digit numbers greater than 5132. You will win for any number greater than or equal to 10,000 and less than or equal to 49,999. Counting the end points, this interval has 40,000 numbers. You will also win for numbers greater than or equal to 100,000 and less than or equal to 260,350 (the greatest number in the list). How many numbers are there in this interval?

b. What is the total amount of numbers in the two intervals in part **a**? Theoretically, these are the numbers on which you can win.

c. Your opponent will win on numbers greater than or equal to 5132 and less than or equal to 9999; and on numbers greater than or equal to 50,000 and less than or equal to 99,999. What is the total number of numbers in these two intervals?

★ d. You can see from the answers in parts **b** and **c** that the intervals within which you will win are bigger than those of your opponent. According to these numbers, about how many times greater is your chance of winning?

★ e. There are 112 populations in the preceding table. How many of these populations will you win on? How many populations will your opponent win on?

Just for Fun—Cryptanalysis

Cryptanalysis is the analysis and deciphering of cryptograms, codes, and other secret writings. There are many historical accounts of situations where successful cryptanalysis was important in gaining military victories and in preventing crime and espionage. One of the earliest cryptographic systems known was used by Julius Caesar and is called the Caesar Cipher. Enciphering a message using this system consists of replacing each letter of the alphabet by the letter which is three letters beyond: A is replaced by D; B by E; C by F; etc. Use the Caesar Cipher to decipher the statement on this scroll.

It is not known why Caesar selected the number 3 as the amount of shift of the alphabet. Any whole number greater than 1 can be used. Another improvement in cryptograms is the elimination of word lengths. The following message is written in groups of five letters and was enciphered by shifting each letter of the alphabet the same number of letters to the right.

OITQT	MWAPW	EMLUM	VWNAK	QMVKM	BPIBE	MQOPQ	VOIVL	UMIAC
ZQVOI	ZMEWZ	BPEPQ	TMVME	BWVKW	VDQVK	MLITI	ZOMXZ	WXWZB
QWVWN	BPMUB	PIBEM	QOPQV	OIVLU	MIACZ	QVOIZ	MBPMW	VTGQV
DMABQ	OIBQW	VABPI	BIZME	WZBPE	PQTMK	PIZTM	AAQVO	MZ

One method of determining the amount of shift is to make a frequency distribution of the letters in the message. The relative frequencies of the occurrence of our letters in large samples of words vary for different letters. The letter E occurs about 13 percent of the time and the letters K, X, and J occur only half of 1 percent of the time. The follow-

A	B	C	D	E	F	G	H	I	J	K	L	M	N	O	P	Q	R	S	T	U	V	W	X	Y	Z
‡	−	≡	≡	‡	≡	−	−	≡	‡		≡	≡	‡	‡	≡		‡	‡	‡	≡	−	−	≡		≡
≡				‡			≡						≡	≡			≡	≡	≡						
				≡															‡						

A	B	C	D	E	F	G	H	I	J	K	L	M	N	O	P	Q	R	S	T	U	V	W	X	Y	Z

ing tallies represent the percentages (rounded off to the nearest whole number) of the occurrences of letters. There are some important patterns. Among the high-frequency letters we see that A, E, and I are four letters apart; two are next to each other, N and O; and three are side by side, R, S, and T. Among the low-frequency letters, two occur as a pair, J and K, and six are side by side, U, V, W, X, Y, and Z. Mark tallies under the letters A through Z for the frequencies of the letters in the preceding message. You will be able to predict the amount of shift by comparing the two sets of tallies and by looking for patterns of high- and low-frequency letters. Try your prediction by deciphering a few words. When you have found the correct amount of shift, decipher the message.*

---

*For further techniques in cryptography, see A. Sinkov, *Elementary Cryptanalysis* (Washington: Mathematical Association of America, 1966).

Answers to Selected Activities

ACTIVITY SET 1.1

1. b. $1 + 3 + 5 + 7 + \cdots + \square = n^2$
2. b. $1 + 2 + 3 + \cdots + n + \cdots + 3 + 2 + 1 = n^2$
3. b. $1 + 2 + 3 + 4 = (4 \times 5)/2$ c. $(9 \times 10)/2$ d. $(100 \times 101)/2$
   e. $1 + 2 + 3 + 4 + \cdots + n = \dfrac{n \times (n + 1)}{2}$
4. Yes. $8 + 16 + 24 + 32 + 40 + 48 + 56 = 224$
5. $4 = 2^2$, $4 + 12 = 4^2$, $4 + 12 + 20 = 6^2$,
   $8^2 = 4 + 12 + 20 + 28$, $10^2 = 4 + 12 + 20 + 28 + 36$
6. Here are two possible patterns.

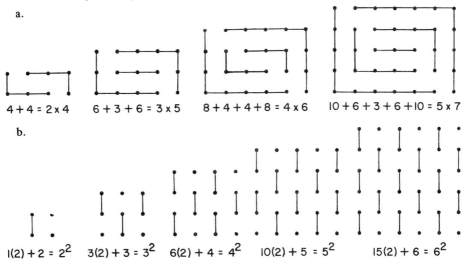

a.

$4 + 4 = 2 \times 4$ $6 + 3 + 6 = 3 \times 5$ $8 + 4 + 4 + 8 = 4 \times 6$ $10 + 6 + 3 + 6 + 10 = 5 \times 7$

b.

$1(2) + 2 = 2^2$ $3(2) + 3 = 3^2$ $6(2) + 4 = 4^2$ $10(2) + 5 = 5^2$ $15(2) + 6 = 6^2$

1.1 Just for Fun—Beginning with the sequence, "She loves me, she loves me not," etc., the last response will be "she loves me," for daisies with an odd number of petals.

1.2 Just for Fun—The winning strategy on a 3 by 3 or 5 by 5 grid is to occupy the center hexagon on the first move. On a 4 by 4 grid the winning first move is any of those which are marked at the right.

4 by 4

ACTIVITY SET 1.3

1. 1 disc, 1 move; 2 discs, 3 moves; 3 discs, 7 moves; 4 discs, 15 moves; 5 discs, 31 moves
4. b. The number of moves for 20 discs is twice the number of moves for 19 discs, plus 1.
5. a. $2^{64} - 1$
6. a.

Number of Discs	1	2	3	4	5	6	7	8	9	10	11	12	13	14	15
Number of Moves	1	3	5	9	13	17	25	33	41	49	65	81	97	113	129

b. 161

ACTIVITY SET 2.1

1. a. *BH* and *LR*; *BH* and *SS*; *BH* and *SR*; *LR* and *SR*; *LR* and *SS* b. *LR* and *SS*
   c. $LYR \in LR$, $SBS \in SS$, $SBH \in BH$
2. a. No b. No
3. a. $\{SRT, SYT, SBT\}$
   c. $\{LRC, LRT, LRH, LRS, LRR, SRT, SRH, SRS, SRC, SRR, SYT, SBT\}$
4. a. Square
   b. Set *B* contains all square, rectangular, and hexagonal attribute pieces. Set *C* contains all square attribute pieces together with all rectangular attribute pieces. Set *D* contains all triangular attribute pieces.
   c. $B \not\subset A$, $C \subset B$, $D \subset A$, $C \not\subset A$, $A \not\subset B$
5. c. No d. These conditions are not the same; they describe different sets.
6. a. $D = G$ and $E = F$
   b. Not(*A* and *B*) has the same meaning as (not *A* or not *B*). Not(*A* or *B*) has the same meaning as (not *A* and not *B*).
7. a. $X = \{SBS, SYS\}$
   b. $Y = \{\quad\quad\}$
   c. $Z = \{LYH, SRH\}$

ACTIVITY SET 2.2

1. a.

B	F	L	U
1	0	0	2

2. a.

B	F	L	U
0	3	2	4

c.

B	F	L	U
1	0	2	2

e.

B	F	L	U
3	3	0	3

3. a. 329 c. 379

4. a.
B	F	L	U
2	1	2	1

b. If there are three multibase pieces of a given kind, they can be replaced by one multibase piece for the next higher power of 3.

5. a.

6. a. There should be 5 markers, because 1 flat is equal to 5 longs.
   c. 14
7. a. Seven   c. 12
9. a. 1443_{five} c. 10212_{three}

ACTIVITY SET 2.3

1. c. The numbers for the FORWARD command for the horizontal movement vary from -129 to 129, and the numbers for the FORWARD command for the vertical movement vary from -109 to 109. These ranges keep the target on the screen.
2. c. Use B = INT(128 * RND(1)) in line 10. The minimum number of guesses is 7.
3. b. To ensure that no two digits in the 3-digit number will be equal.

ACTIVITY SET 3.1

2. Player 1
3. a. 2012_{five} c. 4034_{five}
5. 1436_{eight} Regrouping is also needed from the longs to the flats.
   $+257_{eight}$
   $\overline{1715_{eight}}$
7. a. 2120_{three} c. 2212_{six}
8. b. 125   c. It is possible to win a block in four turns.

9. a. 512   b. 15
10. a. 5204_{seven} c. 3522_{six}

ACTIVITY SET 3.2

1. The player has 4 flats, 1 long, and 2 units after the first turn.

2. Player 1

B	F	L	U
	3	2	3
		4	0
	2	3	3

Player 3

B	F	L	U
	3	1	4
	1	2	1
	1	4	3

3. a. 424_{five} c. 3414_{five}

4. Player 1

B	F	L	U
	6	6	1
		2	6
	6	3	2

Player 3

B	F	L	U
	6	4	1
		5	1
	5	6	0

6. Player 1

B	F	L	U
1	0	0	0
		1	8
	9	8	2

7. a. 1222_{three} c. 1514_{eight}

9. a. To compute this difference, we first decrease the 7 in the tens column to a 6 and add 2 to the 1 in the units column. The remaining steps of the algorithm are carried out as usual. This algorithm has the advantage that borrowing is eliminated and replaced by differences from 10 (10 − 8, 10 − 3, etc.).

```
    6
  34⁷1
  −128
```

ACTIVITY SET 3.3

1. a. 3023_{five} c. 14776_{eight}
2. a. $10_{five} \times 1234_{five} = 12340_{five}$ $10_{five} \times 203_{five} = 2030_{five}$
   b. Place a "0" at the right end of the numeral.

c.

Interval	1 to 40	40 to 80	80 to 120	120 to 160	160 to 200
Number of Primes	12	10	8	7	9

3. Keith needed 11 colors for his grid.
   b. The numbers that have both 2 and 7 as factors are in boxes with a black and a blue corner. (These boxes may also have other colors.) These numbers are 14, 28, 42, 56, 70, 84, 98, and 112.
   d. The numbers which are multiples of both 3 and 5 are in boxes which have a red corner and a green corner. (These boxes may also have other colors.) These numbers are 15, 30, 45, 60, 75, 90, 105, and 120.
4. a. The multiples of 5 lie along parallel lines with four vertical spaces between adjacent lines.
   c. As the primes get greater the pattern of lines gets steeper.
   e. Except for 2 and 3, the primes are either 1 less than a multiple of 6 (in the 5th column) or 1 more than a multiple of 6 (in the 1st column). g. No
5. a. Every other diagonal contains even numbers, and the diagonals between these contain odd numbers. Except for 2, all primes are contained in every other diagonal.
   d. Five: 79, 47, 23, 7, and 19

3.5 Just for Fun—Spirolaterals b and e cannot be completed. The first and third of the following E S W N tables show that the spirolaterals can be completed because the sums in the East-West directions are equal and the sums in the North-South directions are equal. The spirolateral in the second table cannot be completed.

a. 1,2,3

E	S	W	N
1	2	3	1
2	3	1	2
3	1	2	3
6	6	6	6

b. 1,2,3,4

E	S	W	N
1	2	3	4
1	2	3	4
1	2	3	4
3	6	9	12

c. 1,2,3,4,5,6

E	S	W	N
1	2	3	4
5	6	1	2
3	4	5	6
9	12	9	12

The least common multiple of 4 and any odd number is four times the odd number. For example, l.c.m. (4,7) = 4 × 7 or 28. Therefore, if the order of a spirolateral is odd, the sequence of numbers will be needed four times before the beginning number and the beginning direction will be together again. For the spirolateral 1,2,3,4,5,6,7 the numbers are needed 28 times before the East direction and the number 1 are together again.

ACTIVITY SET 3.6

1. a. 6 and 3   c. No, each square number, 1, 4, 9, 16, etc., has an odd number of factors.
2. c. Both 2 and 3 are not factors of the number which is represented by this train.
3. a. The 3 yellow rods show that both 3 and 5 are factors of the number which is represented by the train, and the 5 green rods show that both 5 and 3 are factors of

this number. Since this is the smallest such train, the number represented by the train is the least common multiple of both 3 and 5.

4. d. The all-green trains show that 3 is a factor of one number but not the other. The all-yellow trains show that 5 is a factor of one number but not the other. Since the white rods are the longest rods that can be used to form these single-color trains, the greatest common factor of the two numbers is 1.

5. a. Here are three towers for 72: 1 brown rod and 1 blue; 1 red, 1 purple, and 1 blue; 3 red rods and 2 green rods.

6. b. The red, green, yellow, and black rods represent the prime numbers 2, 3, 5, and 7. Therefore, the rods in any tower can be replaced by combinations of these 4 rods.

8. a. The black and yellow rods are common to both towers. Therefore, the greatest common factor of 105 and 245 is 7 × 5 or 35.

9. a. Black, green, red, and yellow rods are contained in these towers, and the maximum number of times each rod occurs in a tower is once. Therefore, the least common multiple of 42 and 105 is 7 × 3 × 2 × 5 or 210.

3.6 Just for Fun—

1. Yes   2. Star (n,s) is congruent to star (n,r).
3. Star $(10,4)$ has 2 paths. Star $(6,3)$ has 3 paths.   4. g.c.f. $(n,s) = 1$
5. Star $(9,2)$ has 1 path and star $(14,4)$ has 2 paths. Each path requires 2 orbits.
6. b. $\dfrac{\text{l.c.m.}(n,s)}{n}$ 7. b. 15

ACTIVITY SET 4.1

1. a. Yes   c. Nonagon (9-sided polygon)
2. b.

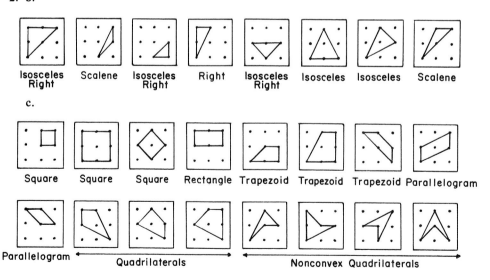

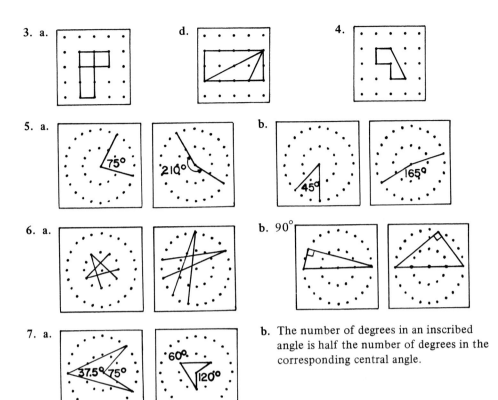

ACTIVITY SET 4.2

1. **b.** There are 120° in each angle of a regular hexagon. Regular hexagons tessellate because sums of combinations of these angles equal 360°.

2. **b.** The number of degrees in an angle of a regular polygon must be a factor of 360.
   c. Each angle has more than 120° (and less than 180°) and so the measures of these angles cannot be factors of 360.

3. **c.** Three. Each has 120°.

4. **b.** Angles *CBA*, *CDE*, and *DEF* meet at one vertex; and angles *BCD*, *BAF*, and *AFE* meet at the other vertex.

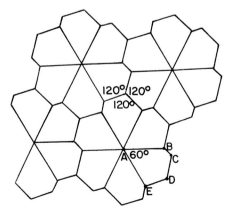
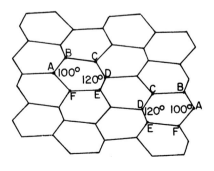

5. a. Semiregular tessellations with two different polygons.

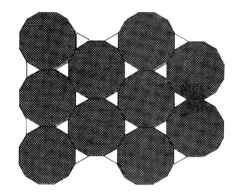

b. Semiregular tessellations with three different polygons.

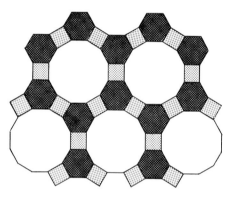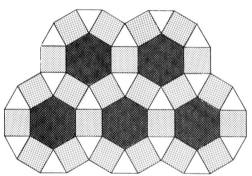

c. 60°, 135°, and 165°

ACTIVITY SET 4.3

1. a. Tetrahedron, 4 triangles; cube, 6 squares; octahedron, 8 triangles; dodecahedron, 12 pentagons; and icosahedron, 20 triangles

2.

	Vertices	Faces	Edges
Tetrahedron	4	4	6
Cube	8	6	12
Octahedron	6	8	12
Dodecahedron	20	12	30
Icosahedron	12	20	30

3. a. (2) c. (5)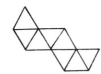

4. b. Octahedron e. Icosahedron f. Cube h. Dodecahedron
   j. Cube or octahedron

5. a. Each vertex is not surrounded by the same number of triangles. Five triangles meet at the top and bottom vertices, and four triangles meet at each of the other vertices.
   c. $3n/2$ is a whole number because it is the number of edges. Since 2 is not a factor of 3, it must be a factor of n.

ACTIVITY SET 4.4

1. a. Angle ABC has a measure of $135°$, $AD = DC$, and $AB = BC$.

    Angle EFG has a measure of $120°$, $EH = HG$, and $EF = 1/2 \times FG$.

2. a. Square with four lines of symmetry.   c. Heptagon with one line of symmetry.
   e. X-shaped figure with one line of symmetry.   f. Nonconvex dodecagon in the shape of a cross with one line of symmetry.

3. a. Regular octagon   e. The eight-pointed wind rose
   f. A regular 16-sided polygon can be obtained by cutting a $157.5°$ angle as shown here, with $AB = AC = AD$. (*Note:* Some of the lines of symmetry are not crease lines.)

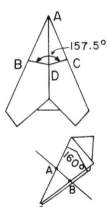

4. A cut along $\overline{AB}$ produces a regular dodecagon with 12 lines of symmetry. A regular 18-sided polygon can be obtained by cutting a $160°$ angle as shown in the accompanying figure. One side of this angle is perpendicular to the edge of the figure and half as long as the other side of the angle.

ACTIVITY SET 5.1

2. a. 155 millimeters   b. 19 millimeters
4. a. For most people this distance will be less than 1 meter.
   b. For some people this distance will be approximately 1 meter.
6. a. A liter is a little bit bigger than a quart. A quart contains approximately 946 cm^3.
   c. Approximately 10.85 centimeters
7. One-half of a meter

ACTIVITY SET 5.2

1. a. 9 square units   c. 7 1/2 square units
2. a. 9 square units   c. 10 1/2 square units
3. a. 3 square units   c. 4 square units
4. a. 3 square units   d. 6 square units
5. a. 4 square units   c. 4 square units   e. 1 square unit
6.

7. The area of a parallelogram is the product of the length of a base times the altitude corresponding to the base.
8. The area of a trapezoid is the product of the altitude times the sum of the lengths of the two bases divided by 2.
9. a. 8 square units   c. A polygon with no interior nails has area $\frac{b}{2} - 1$; a polygon with one interior point has area $\frac{b}{2}$; and a polygon with 2 interior nails has area $\frac{b}{2} + 1$. In general, the area of any polygon on the geoboard is $\frac{b}{2} + i - 1$.

10. Here is one polygon which satisfies these six conditions. There is at least one other.

5.2 Just for Fun—The 12 pentominoes have a total area of 60 square units. Therefore, the dimensions of rectangles must be factors of 60. Since some pentominoes have a width of 3 squares, the smallest width for these rectangles is 3 and the greatest length is 20.

Pentomino puzzle with
four missing corner squares

The eighth and final move is the
pentomino marked with X's.

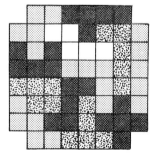 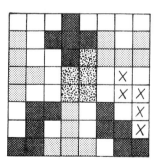

ACTIVITY SET 5.3

1. The volume of the crown was greater than the volume of the gold. King Hiero was cheated.

2. Hexagonal prism: area of base equals 9.1 cm^2; height equals 5.0 cm; volume equals 45.5 cm^3; surface area equals 75.2 cm^2. Cylinder: area of base equals 12.6 cm^2; height equals 3.0 cm; volume equals 37.8 cm^3; surface area equals 62.8 cm^2.

3. The volumes of Cylinder A and Cylinder B are approximately 1028 cm^3 and 1343 cm^3, respectively.

5. b.

Central angle of disc	45°	135°	270°
Base area (cm^2)	4.9	44.2	176.6
Altitude (cm)	9.92	9.27	6.61
Radius (cm)	1.25	3.75	7.50
Volume (cm^3)	16.2	136.4	389.2

  c. As smaller and smaller central angles of less than 45° are used, the volume gets smaller. As larger and larger central angles with more than 315° are used, the volume also gets smaller. The maximum volume occurs for approximately 294°.

6. The "half-cone" has one-eighth the volume of the whole cone. A cone with a base of radius r and a height of h has a volume of $1/3 \times \pi r^2 h$. The cone it contains which is half as high has a base of radius $r/2$ and a height of $h/2$. Its volume is

$$\frac{1}{3} \times \pi \left(\frac{r}{2}\right)^2 \left(\frac{h}{2}\right) = \frac{1}{3} \times \pi r^2 h \times \frac{1}{8}$$

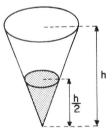

7. a. 7235 cm^3 c. 203.6 cm^3
8. a. Approximately 1/2   c. Slightly less than 1/2

5.3 Just for Fun—The figure with 13 cubes cannot be formed because no combination of 4s and 3 add up to 13.

ACTIVITY SET 6.1

1. a. Four bars have fractions that equal 1/2. Five bars have fractions that equal 0/4.
   b. $\frac{4}{6} = \frac{2}{3} = \frac{8}{12}$, $\frac{3}{6} = \frac{1}{2} = \frac{2}{4} = \frac{6}{12}$
   d. This table has 21 fractions which are not in lowest terms.

2. a. $\frac{1}{6} = \frac{2}{12}$, $\frac{2}{6} = \frac{1}{3}$, $\frac{3}{6} = \frac{1}{2}$, $\frac{4}{6} = \frac{2}{3}$, $\frac{3}{12} = \frac{1}{4}$. There are other possibilities.

3. $\frac{1}{12} < \frac{1}{6} < \frac{1}{4} < \frac{1}{3} < \frac{5}{12} < \frac{1}{2} < \frac{7}{12} < \frac{2}{3} < \frac{3}{4} < \frac{5}{6} < \frac{11}{12}$

4.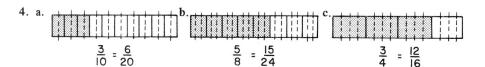

   d. There will be 3a shaded parts out of a total of 3b parts. The fraction for this bar is 3a/3b.

5. b. $\frac{2n}{3n}$ c. It increases both the numerator and denominator by a multiple of n.

6. a.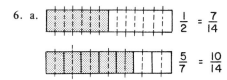

7. a. $\frac{7}{6}, \frac{4}{3}, \frac{3}{2},$ and $\frac{5}{3}$

8. b. $1\frac{1}{8}$ and $1\frac{1}{4}$ d. The blue rod

9. a. The blue rod   c. The green rod

10. a. $\frac{1}{2} = \frac{2}{4}$ c. $\frac{1}{3} = \frac{3}{9}$. There are other possibilities for parts **a** and **c**.

11. a. White and green rods   c. If the unit rod is yellow, the red rod represents 2/5.
    If the unit rod is orange, the purple rod represents 2/5.

ACTIVITY SET 6.2

1. a. $\frac{3}{10} + \frac{2}{5} = \frac{7}{10}$ c. $\frac{1}{3} + \frac{8}{9} = 1\frac{2}{9}$ e. $\frac{3}{8} + \frac{1}{4} = \frac{5}{8}$

3. a.   d. $\frac{5}{6} + \frac{1}{4} = 1\frac{1}{12}$

   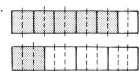   g. $\frac{5}{6} - \frac{1}{4} = \frac{7}{12}$

4. a.    b. 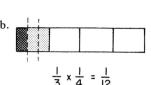   c.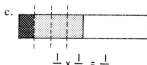

   $\frac{1}{2} \times \frac{1}{3} = \frac{1}{6}$ $\frac{1}{3} \times \frac{1}{4} = \frac{1}{12}$ $\frac{1}{4} \times \frac{1}{2} = \frac{1}{8}$

5. c. 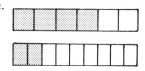 $\frac{4}{6} \div \frac{2}{9} = 3$

6. a. The 32 bars can be paired so that all 16 quotients are whole numbers.

7. a. $\frac{1}{6} + \frac{1}{4} = \frac{5}{12}$ $3 \times \frac{3}{12} = \frac{3}{4}$

$\frac{2}{4} - \frac{1}{2} = \frac{0}{3}$ $\frac{2}{3} \div \frac{2}{6} = 2$

ACTIVITY SET 6.3

1. a. 11   c. Pay 2 yen
3. $^-4 + 1 = {^-3}$

4. a.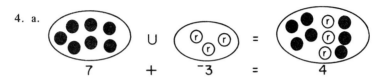

   d. If there are more red chips than black chips, subtract the number of black chips from the number of red chips and make the answer negative. If the number of black chips is greater than or equal to the number of red chips, subtract the number of red chips from the number of black chips.

5. a. $^-3$ c. To subtract a negative number, add its opposite (inverse for addition).

6. a.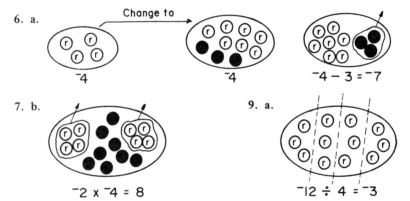

7. b.   9. a.

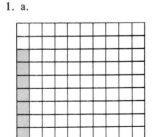

ACTIVITY SET 7.1

1. a.   d. .05
                  2. a. .30   e. 1.0   h. .090

3. a. 3 parts out of 10 is equal to $\boxed{30}$ parts out of 100.
.3 = $\boxed{.30}$

c. 470 parts out of 1000 is equal to $\boxed{47}$ parts out of 100.
.470 = $\boxed{.47}$

e. 1 part out of 100 is equal to $\boxed{10}$ parts out of 1000.
.01 = $\boxed{.010}$

5. a. 7 tenths and 2 hundredths regroup to 6 tenths and $\boxed{12}$ hundredths.

e. 36

6. a. 2 tenths, 6 hundredths, and 15 thousandths regroup to 2 tenths, $\boxed{7}$ hundredths, and 5 thousandths.

7. b. There will be 10,000 parts, and the decimal for one of these parts is .0001.

8. a. .600   The greatest decimal is .62
   c. ⓪.0420   .0047   .0400

9. b. The hundredths square should have 33 shaded parts, and the thousandths square should have 333 shaded parts.

ACTIVITY SET 7.2

1. d. Place one decimal under the other so that the decimal points are aligned, and add the two numbers as though they were whole numbers. Then place the decimal point in the sum directly below the decimal points in the summands.

3. b. If there were 4 decimal squares for .725, there would be a total of $\boxed{2900}$ shaded parts. Since there are 1000 parts in each whole square, this equals $\boxed{2}$ whole squares and $\boxed{900}$ parts out of 1000.

4. a. If there were 10 decimal squares for .572, there would be a total of $\boxed{5720}$ shaded parts. Since there are 1000 parts in each whole square, this equals $\boxed{5}$ whole squares and $\boxed{720}$ parts out of 1000.
   c. To multiply by 10^n, move the decimal point n places to the right.

5. a.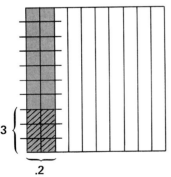

   d. The product .06 × .01 means "take $\boxed{.06}$ of $\boxed{.01}$." To do this we can split the shaded amount of a $\boxed{.01}$ square into $\boxed{100}$ equal parts and take $\boxed{6}$ of them. Since each new part is ten-thousandths of a whole square, .06 times .01 is equal to $\boxed{.0006}$.

6. a.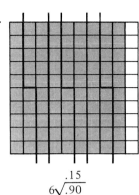

$6\sqrt{.90}$ with quotient .15

e. Use long division to divide as though you were dividing one whole number by another. Then place the decimal point for the quotient directly above the decimal point in the dividend.

7. a. Replace a .3 square by a .30 square. Partition the shaded amount of the decimal square for .30 into 10 equal parts. One of these parts is .03 of a whole square.

8. b.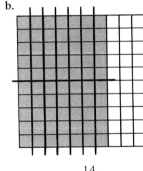

$.05\sqrt{.70}$ with quotient 14.

d. *Measurement Concept:* .2 "fits into" .9 four times with .1 left over. The ratio of the remainder .1 to the divisor .2 is 1 to 2 or .5. So, $.9 \div .2 = 4.5$.

Partitive Concept: Since $.9 \div .2 = 9 \div 2$ and $9 = 8 + 1$, first divide 8 by 2 to get 4. Then replace 1 by 10 tenths and divide by 2 to get 5 tenths or .5. So, $9 \div 2 = 4.5$.

ACTIVITY SET 7.3

1. b. Rather than subtract 22% of 14.85 from 14.85, the same result can be obtained by computing 78% of 14.85. This equals $11.58. Distributive property.

2. b. Using the distributive property, $.06 \times 6.28 + 6.28 = (.06 + 1) \times 6.28$, and $.06 + 1 = 1.06$.

3. b. 12   4. b. Compounding quarterly; 41 cents

5. a. Enter $\boxed{\_\_4\,7\,0\,0\,0}$ b. Enter $\boxed{.0\,0\,0\,0\,8\,7}$

   Scientific notation $\boxed{4.7\_\_\_0\,4}$ Scientific notation $\boxed{8.7\_\_\_^-0\,5}$

   d. 1.0×10^{-99} which equals $\frac{1}{10^{99}}$

6. a. |1.0 _ _ _ _ 1 2| c. |2.7 2 _ _ _⁻1 2|

7. c. Multiply by 10 and subtract 5.

8. b. $\frac{1}{2}+\frac{1}{4}+\frac{1}{8}+\frac{1}{16}+\frac{1}{32}+\frac{1}{64}+\frac{1}{128}+\frac{1}{256}+\frac{1}{512} \approx .9980469$. If you continue adding fractions from this sequence, the sum will always be less than 1.

9. b. $1472081 \div 3876 \approx 379.79386$. The product of .79386 and 3876 equals 3077.0014 which rounded off to the nearest whole number equals the whole number remainder of 3077.

ACTIVITY SET 7.4

1. b.

Area of square	1	2	4	5	8	9	10	16
Length of side	1	$\sqrt{2}$	2	$\sqrt{5}$	$\sqrt{8}$	3	$\sqrt{10}$	4

2. a., b. The A, B, and C squares for the smallest triangle have areas of 9, 4, and 13. The A, B, and C squares for the largest triangle have areas of 20, 45, and 65.

3. a.   c.   f.   i.   j.

4. a. $2\sqrt{2} > 5$ c. $\sqrt{2} + \sqrt{8} > \sqrt{10}$ e. $3\sqrt{2} < 2\sqrt{5}$

5.

Polygon	a	c	e
Perimeter	$2 + \sqrt{5} + \sqrt{8} + \sqrt{13}$	$2 + \sqrt{8} + \sqrt{2} + \sqrt{18}$	$4\sqrt{5}$
Approximate perimeter	10.6	10.4	8.8

6. a. 12 square units   c. 6 square units   e. 8.5 square units

7. Two line segments on a 6 by 6 array have the same length of 5 units. One of these is the hypotenuse of a 3-4-5 right triangle.

ACTIVITY SET 8.1

1. b. $x^2 + 2xy + y^2$

2. a.

(3y)(3y) = 9y²

c.

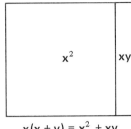

x(x + y) = x² + xy

3. a.

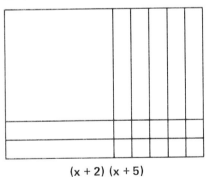

y(2x + y) = 2yx + y²

4. a. A rectangle cannot be formed.
   c. A rectangle can be formed.
5. a. A rectangle can be formed.
   c. A rectangle cannot be formed.
6. a. $(x + 2)(x + 4)$ d. $(x + 1)(2x + 3)$
7. a.

d. Neither a square nor a rectangle can be formed.

(x + 2)(x + 5)

ACTIVITY SET 8.2

1. a. 18 centimeters
   c. The ellipse becomes more circular. The major axis has a length of 22 centimeters, and the length of the minor axis is approximately 21.6 centimeters.
2. b. The distances from the pencil to F_1 and F_2 become smaller, but the difference between these distances will be approximately 6 centimeters.
   c. Both distances are being decreased by the same amount as the string is being pulled.

3. a. The lower edge of the paper
4. a. The center of the circle.   c. The ellipse becomes more circular.
5. a. The center of the circle
6. Circle   7. Hyperbola   9. a. Parabola   d. Hyperbola

ACTIVITY SET 8.3

1. b. $N + 4(1 + 2 + 3 + \ldots + N - 1)$
2. c. 550   e. $\dfrac{N+1}{2N} \times N^3$
3. b. | Figure 1 | Figure 2 | Figure 3 | Figure 4 |
      |---|---|---|---|
      | 5×1^2 | 7×2^2 | 9×3^2 | 11×4^2 |

   d. Figure 4 $4 \times 5 \times 12 - 4^3 = 176$
      Figure 5 $5 \times 6 \times 15 - 5^3 = 325$

5. d. 2550
6. b. | Figure 1 | Figure 2 | Figure 3 | Figure 4 | Figure 5 |
      |---|---|---|---|---|
      | $2 + 3 + 1$ | $4 + 4 + 3$ | $6 + 5 + 6$ | $8 + 6 + 10$ | $10 + 7 + 15$ |

ACTIVITY SET 9.1

1. a., b.

Rotation of 360°   Rotation of 120°   Rotation of 240°   Reflection about L_1 Reflection about L_2 Reflection about L_3

   d. In 10 different ways (5 rotations and 5 reflections)
2. a. Reflection about the vertical center line of the rectangle
   c. Reflection about the horizontal center line of the rectangle and then a translation of the image two boxes to the left

4. c.    5. b.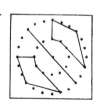

6. a. 120° clockwise rotation   d. 180° rotation about the intersection of lines R and S.

ACTIVITY SET 9.2

1. a. Both triangles contain the angle at E, and the angles at A and A' are right angles. Therefore, the corresponding angles of the two triangles have the same numbers of degrees.
2. c. Triangle E_1FE_2 is similar to triangle RFT. Therefore,
$$\frac{75}{7.5} = \frac{FT}{90} \quad \text{or} \quad FT = 10 \times 90$$
   e. For estimating large distances the distance from eye to finger is relatively small, and ignoring it will not significantly affect the estimate. However, for small distances which may only be a few times greater than the distance from eye to finger, this distance should be added to the estimated distance.
3. b. 20.25 meters   c. 24.3 meters
4. a. ET in the figure equals 80 millimeters. Since this distance represents 60 meters, one possible scale is 1 millimeter for each .75 meter. The height of the tree from T to R is $.75 \times 27 = 20.25$ or approximately 20 meters.
   d. Approximately 19 meters
5. a. Approximately 33 meters
6. b. Use the line from the land marker to the pine tree as a base line and measure off this distance. Measure the five angles from the pine tree to the remaining 5 objects, and then measure the 5 angles from the land marker to the 5 objects.

ACTIVITY SET 9.3

1. a. Bunyan   b. Two sided
3. a. Two bands; the smaller one is one sided and the larger one is two sided.
5. There will always be a winner.
6. Slip the string connecting one person's wrists under the string which is around the wrist of the second person and over this person's hand.
7. a. The greatest number of turns is 5 and the least number is 4.
   c. The greatest number of moves is $3n - 1$ and the least number of moves is $2n$. (*Note:* In some cases the least number may appear to be less than $2n$. Remember, an arc can be drawn from a point back to itself.)
8. *Hint:* Pull the strings through the hole of the panel.
10. No. If the arms are intertwined so that a knot is formed, this cannot be changed into a simple closed curve without breaking one or more hand grips.

ACTIVITY SET 10.1

2. d. $\frac{1}{4}$ e. No
3. b. $\frac{4}{9}$ c. $\frac{9}{16}$
4. a. Two   c. A_1 A_2 A e. No
           A_2 A_1 B

10.1 Just for Fun—

c. Lazzerini's probability for this experiment is .6364436, and the corresponding value for π is 3.1424619.

e. $\dfrac{1}{\pi}$

ACTIVITY SET 10.2

1. b. $\dfrac{3}{10}$ d. $\dfrac{3}{5}$ 2. b. 6   c. $\dfrac{3}{8}$

3. $\dfrac{7}{15}$ 4. $\dfrac{49}{100}$

5. a. $\dfrac{1}{36}$ c. $\left(\dfrac{35}{36}\right)^2$ e. Approximately 49%

10.2 Just for Fun—

b. $\dfrac{2}{3}$ c. $\dfrac{1}{3}$ d. $\dfrac{1}{3}$ e. $\dfrac{2}{3}$

ACTIVITY SET 10.3

2. The standard deviation is approximately 1.7.
   a. 18 or 75%   b. 22 or approximately 92%

4. b. The greatest possible average is $18,000.00 and the smallest possible average is $9,333.33   d. The greatest possible average is $16,833.33 and the smallest possible possible average is $10,333.33.

ACTIVITY SET 10.4

1. 1 2 3 4   1 2 3 4   1 2 3 4
    Ⓗ Ⓗ Ⓣ Ⓣ   Ⓗ Ⓣ Ⓗ Ⓣ   Ⓗ Ⓣ Ⓣ Ⓗ
    Ⓣ Ⓣ Ⓗ Ⓗ   Ⓣ Ⓗ Ⓗ Ⓣ   Ⓣ Ⓗ Ⓣ Ⓗ

2. b. 68.3%   c. 95%

4. a.

e	s	c	w	k
16.3%	7.1%	3.1%	2%	.6%

c. This assignment of code symbols requires 43 time units as compared to 47 time units for Morse code.

e	t	a	o	n	i	s	r	h
•	—	• •	• —	— •	• • •	— —	• — •	• • • •

5. a. 160,351   d. About 3.65 times greater   e. You win on 73 populations and your opponent wins on 39 populations.

Index

Abacus:
  base ten, 43–45, 48–51
  division, 48–51
  historical, 23
  multiplication, 41–45
  variable base, 23–25, 42, 50
Acute angle, 163
Addition:
  algorithm, 32–37, 131–132
  decimals, 131–132, 146
  fractions, 112–115
  integers, 117, 119–121
  irrational numbers, 145–146
  models, 32–37, 60–62, 112, 115, 117, 119–120, 131–132
  regrouping, 32–37, 126–128
  signed numbers, 117, 119–121
  whole numbers, 32–37
Aichele, Douglas B., 8
Algebra, 150–155, 167
Algorithm (see Addition, Subtraction, Multiplication, and Division)
Alidade, 180
Altitude:
  cone, 101
  parallelogram, 97
  prism, 100
  trapezoid, 97
  triangle, 96
Aman, George, 97
Ando, Masue, 41
Angle:
  acute, 162
  central, 72–73, 101
  congruent, 17, 72
  elevation, 177–178
  horizontal, 179
  inscribed, 72–73
  obtuse, 162
  measurement, 72–73, 76–79
  right, 18, 73, 96, 145, 176–177

vertical, 179
Angle of elevation, 177–178
Approximation, 146, 174–179
Arabian Nights Mystery, 52
Arc:
  circle, 72–73
  network, 185
Archimedes, 84, 99
Archimedean solids, 84
Area, 95–98, 100–101, 143–144, 146–147
Ashlock, Robert B., 123
Attribute blocks, 16–20
Attribute game, 20
Attribute pieces, 16–20
Average, 198
  mean, 198–202
  median, 198
  mode, 198

Bar Graph, 203
Base:
  parallelogram, 97
  trapezoid, 97
  triangle, 96
Base area, 100–101
Base (numeration systems):
  eight, 34, 36, 39, 42, 46–47, 50
  five, 20–25, 32–33, 35–37, 42, 46, 50
  four, 35, 39, 42, 47
  nine, 47, 50
  seven, 24–25, 37–39, 50
  six, 35, 37, 50
  ten, 34–35, 39–41, 43–46, 48–51, 123–128
  three, 22, 25, 35, 37, 39, 46–47, 50
Base ten abacus, 43–45, 48–51
BASIC programs, 27–28
Battleships (game), 162–163
Bedford, Fred L., 175
Bell-shaped curve, 201
Bell, William R., 162

Bennett, Albert B., Jr., 64, 117
Binary numbers, 28–30
Black and red chips model, 116–119
Bohan, Harry, 107
Borrowing, 41 (see also Regrouping)
Boundary point, 97–98
Bradford, Cheryl L., 56
Brahmans, 11–12
Buchman, Aaron L., 144
Buffon, Count, 182
Buffon's Needle Problem, 192–193
Bunyan, Paul, 181–182

Caesar Cipher, 207
Caesar, Julius, 207
Calculator:
  cross-number puzzle, 47
  decimals, 137–141
  discounts, 137
  fractions, 140–141
  game, 52
  hidden digits, 140
  interest rates, 138–139
  number tricks, 52
  percent, 137–139
  sales tax, 138
  scientific notation, 139–141
  whole numbers, 47–48, 52–53
Card and Ring Puzzle, 185
Carpenter, Thomas P., 111, 123, 131
Carrying (see Regrouping)
Carson, George S., 103
Cartesian coordinate system, 162–163, 157–158
Centimeter, 91–94
Centimeter and Meter Guessing Game, 94
Centimeter Racing Game, 94
Central angle, 72–73, 101
Characteristic, 139–140

231

Chinese:
  game of nim, 30
  model for signed numbers, 116
  tangram puzzles, 74
Chip Trading, 35–37, 46–47
Chi square test, 204
Circle, 26, 156, 160–161
  arc, 72–73
  area, 100–101
  circumference, 100
  diameter, 101
  radius, 101, 160
Circular geoboard, 69, 72–73, 172–173
Circumference, 100
Clemens, Stanley R., 81
Clinometer, 177
Closed curve, 186
Codes, 105, 207
Commensurable, 143
Common denominator, 108, 112
Complement of a set, 18–19
Complmentary probability, 196
Composite number, 55
Composition of mappings, 172–173
Compound probability, 193–187
Computer Gamess, 25–28
Cone, 101, 156
  altitude, 101
  base area, 101
  radius, 101
  volume, 101–102
Congruent:
  angles, 17, 72
  girures, 172
  line segments, 17
Conic sections:
  circle, 156, 160–161
  ellipse, 156–157, 159, 161
  hyperbola, 156, 158–159, 161
  parabola, 156, 158–159, 161–162
Conjecture, 6, 59–60, 65–66, 97, 173, 182, 186
Continuous path, 65–67
Convex, 69–70, 81, 85
Corbitt, Mary K., 111, 123, 131
Corresponding sides, 174
Courant, Richard, 143

Cross-number puzzle, 47
Cruikshank, Douglas E., 20
Cryptanalysis, 207,–208
Cryptogram, 207
Cube, 82–83, 160
Cubic centimeter, 91, 93, 100–102
Cuisenaire rods, 60–63, 105, 108–109
Curiosities of the Cube, 83
Curve:
  circle, 26, 156, 160–161
  closed, 186
  ellipse, 156–157, 159, 161
  hyperbola, 156, 158–159, 161
  normal, 201–203
  parabola, 156, 158–159, 161–162
  spiral, 5–6, 57
Cylinder:
  circumference, 100–101
  base area, 100–101
  diameter, 100–101
  volume, 100–101

Davidson, Patricia S., 32
DaVinci, Leonardo, 81
Decimal Bingo, 130
Decimal Place Value Game, 131
Decimal Square, 123–136
Decimal Squares Black Jack, 136
Decimals:
  addition, 131–132, 146
  approximation, 129–130, 146
  calculator, 137–141
  converting from fractions, 140–142
  converting to fractions, 142
  division, 134–136
  equality, 124–125
  games, 130–131, 136
  inequality, 128–129
  infinite nonrepeating, 143, 148
  infinite repeating, 141
  models, 123–130
  multiplication, 132–134
  part-to-whole concept, 123–124
  place value, 125–126, 131
  regrouping, 126–128
  subtraction, 132
Decimeter, 91
Decipher, 207–208

Decomposition method, 41
Deductive reasoning, 6, 8–9, 27–28
Degree, angle:
  central, 72–73, 101
  elevation, 177–178
  inscribed, 72–73
  polygon, 78
  rotation, 170–173
  tessellation, 76–78
  triangle, 76
Deltahedra, 85
Denominator, 106–108
Dependent event, 193–194
Dependent variable, 167
Diameter, 101
Dienes, Zoltan P., 16, 20, 170
Directrix, 158
Disc, 101
Discounts, 137
Disjoint sets, 16–17
Distribution:
  even, 205
  normal, 201–203
  random, 203–206
  uneven, 204
  uniform, 201, 203–204
Division:
  algorithm, 48–52, 134–135
  decimals, 134–135
  fractions, 113–115
  integers, 119, 121
  long division, 48–52, 134–135
  measurement, 48–51, 119, 135
  models, 48–52, 113–115, 119, 134–135
  partitive, 119, 134
  powers of ten, 135
  quotient, 49–51, 113–115, 134–135, 141
  regrouping, 49–51
  signed numbers, 119, 121
  whole numbers, 48–51
Dodecahedron, 82–83
Domino, 99
Driscoll, Mark J., 105, 131
Drizigacker, Rowena, 110
Duncan, David R., 79

Egyptians, 81
Elementary Cryptanalysis, 208
Element of, 16
Elevation, angle of, 177–178
Ellipse, 156–157, 159, 161

232 INDEX

Empirical probability, 189–197
Empty set, 19
Encipher, 207
Envelope:
  ellipse, 159
  hyperbola, 159
  parabola, 158
Equal:
  fractions, 105–110
  sets, 18–19
Equation:
  fractions, 105–109, 112–114
  integers, 117–121
  irrational numbers, 145–146
  whole numbers, 2–5
Equilateral triangle, 16, 76, 78, 85–86
Equivalent sets, 16–17
Eratosthenes, 53, 55
Error Patterns in Computation: A Semi-Programmed Approach, 123
Escher, Maurits C., 75, 80–81
Estimation, 92, 94
Even number, 37
Event:
  compound, 193–197
  dependent, 193–194
  independent, 193
Eves, Howard W., 41
Existence proof, 10

Factors, 56, 60–64
Fair, Arlene W., 32
Fair game, 190
Fechner, Gustav, 147
Fibonacci numbers, 5–6, 148
Field Work in Mathematics, 175
Five-cornered toy, 171
Five pointed star, 89
Flipping motion (*see* Reflection)
Focus, 157–159
Fraction Bar Black Jack, 115
Fraction Bars, 105–108, 111–115
Fraction Bingo, 110
Fractions:
  addition, 112–115
  calculator, 140–141
  common denominator, 107–108, 112
  converting to decimals, 140–142
  denominator, 105–108, 112
  division, 113–114
  equality, 105–110
  improper, 108
  inequality, 106, 110
  lowest terms, 106
  mixed numbers, 108
  models, 105–110
  multiplication, 113–114
  numerator, 105, 108
  subtraction, 112–114
  unit, 109
FRIO (Fractions in Order), 110
Frequency, 198, 201
Frequency distribution, 201–208
Function, 167

Gadsby, 204
Gale Game, 184
Galilei, Galileo, 93
Galton, Grace K., 32
Games:
  Attribute Game, 20
  Battleships, 162–163
  Centimeter and Meter Guessing Games, 94
  Centimeter Racing Game, 94
  Decimal Bingo, 130
  Decimal Place Value Game, 131
  Decimal Squares Black Jack, 136
  Fraction Bar Black Jack, 115
  Fraction Bingo, 110
  FRIO, 110
  Gale Game, 184
  Greatest Quotient, 115
  Grid Game, 20
  Hex, 10
  Hide-a-region, 163
  Hi-Lo (BASIC), 27
  integer games, 119–121
  Keyboard Game, 52
  negative number games, 119–121
  Nim, 30
  Patterns, 6–8
  Pentomino Game, 99
  Pica-Centro, 8, 27–28
  Pick-a-Word, 9
  solitaire games, 114–115
  Sprouts, 185
  Target Game (Logo), 26–27
  Three-Penny Grid, 191
  Trading Down, 37–39
  Trading Up, 32–36
  Two-Penny Grid, 190
  What's My Rule, 167
Games for:
  addition, 32–36, 114–115
  algebra, 155
  area, 163
  binary numbers, 30
  calculators, 52
  computers, 25–28
  coordinates, 162–163
  decimals, 130–131, 136
  division, 114–115
  equality, 110
  fractions, 110, 115
  functions, 167
  functions, 167
  inequality, 110, 27
  integers, 119–121
  metric units, 94
  multiplication, 114, 120–121
  negative numbers, 119–121
  probability, 190–192
  reasoning, 6–9, 27–28
  sets, 20
  subtraction, 37–39, 114–115
  topology, 184–185
Gamut of Games, A, 6
Gardner, Martin, 6, 10, 53, 77, 99, 158, 184–185
Gauss, Karl Friedrich, 2, 3, 60
Geoboard:
  circular, 69, 72–73, 172–173
  rectangular, 69–71, 95–98, 143–147
Geometric patterns, 2–6, 58–60, 64–67, 162, 164–167
Geometry: An Investigative Approach, 81
Goldberg, Kenneth P., 162
Golden Mean, The, 148
Golden ratio, 148
Golden rectangle, 147–148
Golding, E.W., 16
Gram, 91
Graph:
  bar, 203
  frequency distribution, 201–204
  rectangular coordinate, 162–163, 157–158
Graphical analysis, 188
Greater than (>), 145

INDEX 233

Greatest common factor, 60–63
Greatest Quotient (game), 115
Greeks:
 conic sections, 156
 geometric numbers, 150
 golden rectangle, 147–148
Grid Game, 20

Handspan, 91–92
Hein, Piet, 103
Heptagon, 70–71
Hex (game), 10
Hexagon, 10, 16, 70–71, 75–78, 87
Hexominoes, 99
Hidden digits, 140
Hide-a-region (game), 163
Hi-Lo (BASIC), 27
Hipparchus, 174
Hippasus, 143
Holt, Michael, 170
Horizontal, 179
Hyperbola, 156, 158–159, 161
Hypotenuse, 143–145
Hypsometer, 178

Icosahedron, 82–83
Ikeda, Hitoshi, 41
Image, 169, 172–173
Improper fraction, 108
Incommensurable, 143
Independent event, 193
Independent variable, 156
Indirect measurement, 174–180
Inductive reasoning, 6–8
Inequality:
 decimals, 128–129
 fractions, 106, 110
 integers, 120
 irrational numbers, 145
 signed numbers, 120
 whole numbers, 27
Infinite nonrepeating decimals, 143, 148
Infinite repeating decimal, 141
Inscribed angle, 72–73
Instant Insanity (puzzle), 13–14
Integers:
 addition, 117, 119–121
 division, 119, 121
 inequality, 120
 inverses for addition, 116
 models, 116–120
 multiplication, 118–121

 negative, 116–121
 negatives (opposites), 116
 number line, 120
 positive, 116
 subtraction, 117–118, 120–121
Intercepted arc, 72–73
Interest rates, 138–139
Interior point, 97–98
Intersection of sets, 17, 71
Inverses for addition, 116
Irrational numbers, 143, 148
Isometric grid, 60
Isosceles triangle, 70, 175

Jencks, Stanley M., 72

Kasner, Edward, 11
Kepner, Henry S., 111, 123, 131
Keyboard Game, 52
Kilometer, 91, 93
Kline, Morris, 111
Krause, Marina C., 86
Lady Luck, 205
Lazzerini, 193
Learning Logic, Logical Games, 16
Least common multiple, 60–64, 66
Lehrer, Tom, 32
Leibniz, Gottfrid Wilhelm, 137
Length, 91–94
Less than (<), 106, 120, 145
Let's Play Math, 170
Levin, Robert E., 13
Lindquist, Mary M., 111, 123, 131
Line design, 161–162
Line segment, 69, 145, 147
Line of symmetry, 86–89
Linn, Charles F., 148
Liter, 91, 93
Litwiller, Bonnie H., 79
Logo procedures, 26–27
Long division algorithm, 48–52, 134–135
Lowest terms, 106

Magic Formula, 52,
Major axis, 157
Mantissa, 139–140
Mapping:
 reflection, 169–173
 rotation, 169–173

 translation, 169, 171–172
Mathematical Carnival, 185
Mathematical Circles Revisited, 41
Mathematics and Imagination, 11
Mathematics in Western Culture, 174
McGowan, William E., 85
Mean, 198–202
Measurement:
 angle, 72–73, 76–79
 length, 91–94, 144–147
 perimeter, 146
 plane area, 95–98, 143–144, 146–147
 surface area, 100
 time, 93
 volume 91, 95, 99–102
 weight, 91, 99–100
Measurement concept of division, 48–51, 119, 135
Median:
 statistics, 198
 triangle, 85
Meter, 91–94
Metric prefixes, 91
Metric system, 91–94
Metric unit:
 gram, 91
 liter, 91, 93
 meter, 91–94
Midpoint, 85, 160
Mile, 93
Milliliter, 91, 102
Millimeter, 91–92
Mind reading cards, 28–29
Minor axis, 157
Mixed numbers, 109
Mode, 198
Models for algebraic terms, 150–154
Models for equality, 61, 105–110, 116–117, 124–125
Models for numbers:
 decimals, 123–136
 fractions, 105–115
 integers, 116–120
 irrational numbers, 144–147
 signed numbers, 116–120
 whole numbers, 20–25
Models for numeration:
 decimals, 123–126
 fractions, 105–106
 whole numbers, 20–25

Models for operations:
　addition, 32-27, 60-62, 112, 115, 117, 119-120, 131
　division, 48-52, 113-115, 119, 134-135
　multiplication, 42-47, 62-64, 113-114, 118-119, 132-134, 147
　subtraction, 37-41, 112-115, 117-118, 132
Moebius band, 181-183
Morse code, 205
Morse, Samuel, 205
Multibase pieces, 20-22, 32-35, 37-41
Multiple, 53-57, 60-64
Multiplication:
　algorithm, 41-47, 132-134
　decimals, 132-134
　fractions, 113
　integers, 118-121
　models, 42-47, 62-64, 113, 118-119, 132-134, 146-147
　powers of ten, 133
　regrouping, 42-47, 132-133
　signed numbers, 118-121
　whole numbers, 41-47
Musser, Gary L., 117

National Assessment of Educational Progress, 105, 111, 123, 131
Negative numbers, 116-121
Network, 185
Newman, James R., 11, 69, 99
Nim, game of, 30
Noncongruent, 143
Nonconvex, 69-71, 79
Normal curve, 201-203
Normal distribution, 201-203
Null set, 19
Number line, 120
Number theory, 52-67
Numbers:
　binary, 28-30
　composite, 55
　decimal, 123-136
　even, 57
　fractions, 105-115
　integers, 116-121
　irrational, 143-148
　negative, 116-121
　odd, 2, 57
　positive, 116-121

　prime, 55-58, 62-63
　signed, 116-121
　whole, 20-25, 32-67
Numeration, 20-25, 123-126
Numerator, 105

Obata, Gyo, 157
Obtuse angle, 162
Octagon, 70, 87
Octahedron, 82-83
O'Daffer, Phares G., 81
Odd number, 2, 57
Odds, Frank C., 60
One-sided surface, 182-183
Opposites, 116
Orbit, 66-67
Order of Spirolateral, 58-60
Outcome, 189, 192-195, 201
Overholser, Jean S., 163

Paper folding:
　ellipse, 159
　hyperbola, 159
　parabola, 158-159
　reflection, 171
　symmetric figure, 86-89
Parabola, 156, 158-159, 161-162
Parallel, 17, 192
Parallelogram, 70-71, 74, 80, 97
Partitive concept of division, 119, 134
Party trick, 184
Patterns:
　artistic, 64-67, 162
　game, 6-8
　geometric, 2-6, 58-60, 64-67, 161-162, 164-167
　letter frequencies, 204, 208
　number, 2-6, 28-29, 53-58
　spiral, 5-6
　wallpaper and fabric, 169-170
Peck, Donald M., 72
Pendulum, 93
Penetrated tetrahedron, 85-86
Pentagon, 70, 77
Pentominoes:
　game, 99
　puzzle, 98
Percent, 137-139, 142, 204
Perimeter, 146
Perpendicular, 160
Pi (π), 100-101, 192-193

Piaget, Jean, 105
Pica Centro (game), 8
Pick-a-Word (game), 9
Pick's formula, 97-98
Place value:
　decimals, 125-126
　whole numbers, 20-25
Plane area, 95-98, 143-144, 146-147
Plane table, 180
Plato, 81
Platonic solid, 81-83
Polygon, 69-71
　angles, 77-79
　area, 95-98
　perimeter, 146
　regular, 75-79, 82-87, 89
　vertices, 76
Polyhedron, 81-86, 100-102
Polyhedron Models, 82
Polyominoes, 99
Population, 199-200
Positional numeration:
　decimals, 125-127
　whole numbers, 22-25
Positive numbers, 116
Post, Thomas R., 105
Prime factor, 56, 62-63
Prime number, 55-58, 62-63
Prism:
　altitude, 100
　base area, 100
　surface area, 100
　volume, 100
Probability, 188-197
　complementary, 196
　compound, 193-197
　empirical, 189-197
　theoretical, 189-197
Product of mappings, 172-173
Projection point, 180
Proof, existence, 10
Protractor, 101, 179
Puzzle:
　Card and Ring, 185
　Instant Insanity, 13-14
　Pentomino, 98
　Soma Cubes, 103
　Tangram, 74
Pyramid:
　altitude, 100
　base area, 100
　surface area, 100
　volume 100
Pythagorus, 150

INDEX　235

Pythagorean theorem, 143-145

Quadrant, 177
Quadrilateral, 70-71, 97
 parallelogram, 70-71, 74, 80, 97
 rectangle, 16, 70-71, 80, 95-97, 146-148
 rhombus, 74, 87
 square, 16, 70-71, 74, 78, 87, 143-144
 trapezoid, 71, 97, 172
Quart, 93
Quotient, 49-51, 113-115, 134-135, 141

Radius, 101, 160
Random digits, 203-206
Range, 198
Ranucci, Ernest R., 83
Ratio, 174-179
Read, Ronald C., 74
Reasoning:
 deductive, 6, 8-9, 27-28
 inductive, 6-8
Rectangle, 16, 70-71, 80, 95-97, 146-148
Rectangular array, 2-5, 53-57, 150
Rectangular coordinates, 157-158
Rectangular geoboard, 69-71, 95-98, 143-147
Rectangular number, 150
Reflection, 169-173
Regrouping decimals:
 addition, 131-132
 division, 134-135
 multiplication, 132-133
 numeration, 123-126
 subtraction, 132
Regrouping whole numbers:
 addition, 32-37
 division, 49-51
 multiplication, 42-47
 numeration, 21-25
 subtraction, 37-41
Regular polygon, 75-79, 82-87, 89
Regular polyedron, 81-83
Regular tessellation, 76-77
Remainder, 141
Research Within Reach Elementary School Mathematics, 105, 131

Reys, Robert E., 123, 131
Rhombus, 74, 87
Right angle, 18, 73, 96, 145, 176-168
Right triangle, 70, 96, 144-145, 176-177
Robbins, Herbert, 143
Rollins, Wilma E., 83
Roman abacus, 23
Rotation, 169-173
Rotational symmetry, 86
Ruled table, 23

Sackson, Sidney, 6
Sales Tax, 138
Sampling, 199-200
Sampling, 200
Scale, 177-180
Scalene triangle, 70, 96
Scientific notation, 139-141
Scout sighting method, 175
Sector of a circle, 101
Semicircle, 73
Semiregular polyhedron, 84
Semiregular tessellation, 78-79
Sequence, 58, 140, 148, 164-167
Sets:
 attribute pieces, 16-20
 complement, 18-19
 disjoint, 16-17
 element of, 16
 empty, 19
 equal, 18-19
 equivalent, 16-17
 intersection, 17, 71
 null, 19
 subset, 18
 union, 17, 71, 117
 Venn diagram, 17, 20
Sextant, 177
Seymour, Dale G., 162
Sholes, Christofer, 204
Shuster, Carl N., 175
Sieve of Eratosthenes, 55
Sighting method, 174
Signed numbers, 116-121
Similarity mapping, 180
Similar triangles, 174-179
Sinkov, Abraham, 208
Sixth Book of Mathematical Games from Scientific American, 53
Sliding motion (*see* Translation)
Slonim, Morris, 200

Snider, Joyce, 162
Solitaire game, 114-115
Soma Cubes, 103
Soroban, 41
Sphere, 102
Spiral, 5-6, 57
Spirolateral, 58-60
Sprouts (game), 185
Square, 16, 70-71, 74, 78, 87, 143-144
Square number, 150
Square root, 144-147, 199
Stadiascope, 176
Standard deviation, 198-199, 202
Star polygon, 64-67
Statistics, 198-208
Stratified sampling, 200
Subset, 18
Subtraction:
 algorithm, 37-41, 132
 decimals, 132
 decomposition method, 41
 fractions, 112-113
 integers, 117-118, 120-121
 models, 37-41, 112-113, 117-118, 132
 regrouping, 37-41
 signed numbers, 117-118, 120-121
 whole numbers, 37-41
Surface area, 100
Swallow, Kenneth P., 57
Symmetry, 75, 86-89

Tangram puzzle, 74
Tangrams, 74
Tapson, Frank, 57
Target Game (Logo), 26-27
Teacher's Guide for Attribute Games and Problems, 16
Teeters, Joseph L., 80
Tessellation, 75-80
 Escher-type, 80
 nonconvex polygon, 79
 regular, 76-77
 semiregular, 78-79
Tetrahedron, 82-86
Tetrominoes, 99
Theodolite, 179
Theoretical probability, 189-197
Three-cornered toy, 170-171
Three-Penny Grid (game), 191
Tile (*see* Tessellation)

Tower of Brahma, 11–13
Trading Down Game, 37–39
Trading Up Game, 32–36
Transit, 179
Translation, 169, 171–172
Trapezoid, 71, 97, 172
Tree diagram, 195
Triangle:
  altitude, 96
  area, 96, 146–147
  base, 96
  equilateral, 16, 76, 78, 85–86
  isoscles, 70, 175
  median, 85
  right, 70, 96, 144–145, 176–177
  scalene, 70, 96
  similar, 174
Trick dice, 196–197
Tricks:
  Arabian Nights Mystery, 52
  Magic Formula, 52
  Party Trick, 184
  Trick Dice, 196–197
  Vest Trick, 186
Trominoes, 99
Turning motion (*see* Rotation)
Twin primes, 55, 57
Two-Penny Grid (game), 190
Two-sided surface, 182–183

Ulam, Stanislaw M., 58
Uniform distribution, 201, 203–204
Union sets, 17, 71, 117
Unit, 109
Unit of measure:
  area, 95-98, 100, 143–144, 146–147
  length, 91–94, 143
  time, 93
  volume, 91, 93, 100–102
  weight, 91
Upson, William H., 182

Variable, 167
Variable base abacus, 23–25, 42, 50
Venn diagram, 17, 20
Vertex:
  angle, 72
  network, 185
  polygon, 76, 163
  polyhedron, 81, 83–84
  tessellation, 76–79
Vertical, 179
Vest trick, 186
Voltaire, 188
Volume, 91, 93, 99–102

Wadlow, Robert, 198–199

Weaver, Warren, 205
Weight, 91, 99–100
Wenninger, Magnus, J., 82
Weyl, Herman, 75
What Is Mathematics, 143
What's My Rule (game), 167
Whole numbers:
  addition, 32–37
  calculator, 47–48, 52–53
  composite, 55
  division, 48–51
  even, 57
  factor, 56, 60–64
  models, 20–25
  multiple, 53–57, 60–64
  multiplication, 41–47
  numeration, 20–25
  odd, 2, 57
  place value, 20–25
  prime, 55–58, 62–63
  subtraction, 37–41
Wind rose, 86
World of Mathematics, The, 99

x-axis, 157–158

Yard, 92
y-axis, 157–158

Materials Appendix

1. Game of Hex Grids (1.2)
2. Tower of Brahma (1.3)
3. Tower of Brahma Discs (1.3)
4. Hide-a-region and Battleships Grids (8.2)
5. Attribute Game Grid (2.1)
6. Geometric Models for Algebraic Expressions (8.1)
7. Variable Base Abacus (2.2, 3.3, 3.4)
8. Mind Reading Cards (2.3)
9. Rectangular Coordinate System (8.2)
10. Rectangular Coordinate System (8.2)
11. Chip Trading Mat (3.1, 3.3)
12. Two-Penny Grid (10.1)
13. Base Ten Abacus—Whole Numbers (3.3, 3.4)
14. Rectangular Grid (3.5)
15. Isometric Grid (3.5)
16. Three-Penny Grid (10.1)
17. Rectangular Geoboard Template (4.1, 5.2, 7.4)
18. Circular Geoboard Template (4.1, 9.1)
19. Buffon's Needle Problem (10.1)
20. Centimeter Racing Mat (5.1)
21. Hypsometer–Clinometer (9.2)
22. Algebraic Skill Cards (8.1)
23. Polygons for Semiregular Tessellations (4.2)
24. Semiregular Polyhedral Faces (4.3)
25. Attribute Pieces (2.1)
26. Prism, Pyramid, and Cylinder (5.3)
*27. Fraction Bars (6.1, 6.2)
*28. Fraction Bars (6.1, 6.2)
*29. Fraction Bars (6.1, 6.2)
30. Black and Red Chips (6.3); Markers for Abacuses (2.2, 3.3, 3.4, 7.1, 7.2)
31. Base Five Multibase Pieces (2.2, 3.1, 3.2)
*32. Decimal Squares (7.1, 7.2)
*33. Decimal Squares (7.1, 7.2)
*34. Decimal Squares (7.1, 7.2)
*35. Decimal Squares (7.1, 7.2)
*36. Chip Trading Chips (3.1, 3.3)
*37. Cuisenaire Rods (3.6, 6.1)
38. Regular Polyhedra—Cube, Octahedron, Tetrahedron (4.3)
39. Regular Polyhedra—Icosahedron, Dodecahedron (4.3)
40. Deltahedra—Isometric Grid (4.3)
41. Metric Ruler, Protractor, and Compass (4.1, 5.1, 5.3, 8.2, 9.2)
42. Trick Dice (10.2)

*Material Cards 27, 28, and 29 contain copies of colored laminated bars; Material Cards 32, 33, 34, and 35 contain copies of colored squares; and Material Card 36 contains copies of colored plastic chips. These materials are commercially produced by Scott Resources, Fort Collins, Colorado. Material Card 37 contains copies of colored wooden rods commercially produced by Cuisenaire Company of America, New Rochelle, New York.

MATERIAL CARD #1—1.2 JUST FOR FUN

Game of Hex Grids

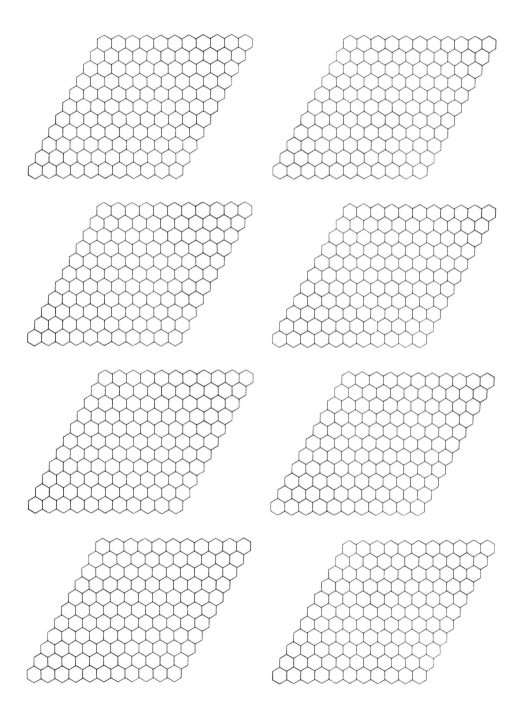

MATERIAL CARD #2 — ACTIVITY SET 1.3

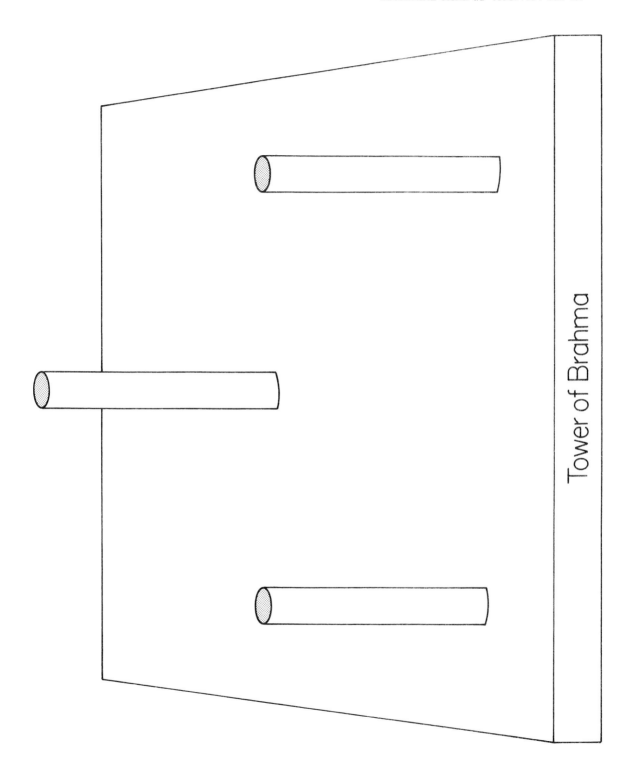

MATERIAL CARD #3—ACTIVITY SET 1.3

Tower of Brahma Discs

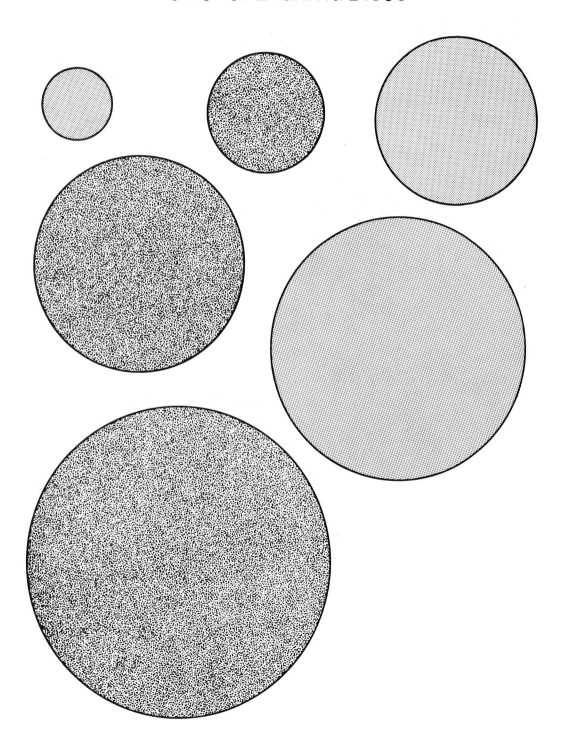

MATERIAL CARD #4—ACTIVITY SET 8.1

Hide-a-region and Battleships Grids

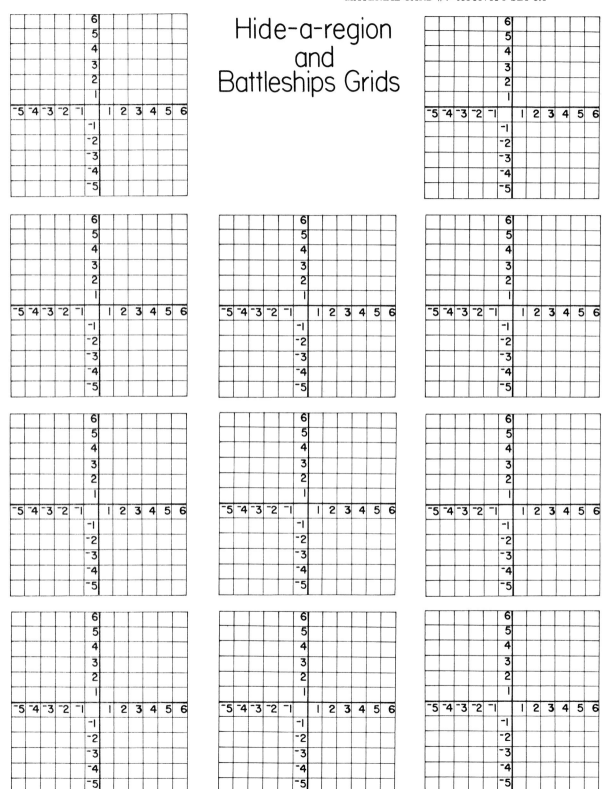

MATERIAL CARD #5—2.1 JUST FOR FUN

Attribute Game Grid

Rows, 1 difference (1 point)   Columns, 2 differences (2 points)   Diagonals, 3 differences (3 points)

		Place attribute piece here to start game		

Geometric Models for Algebraic Expressions

MATERIAL CARD #6 — ACTIVITY SET 8.1

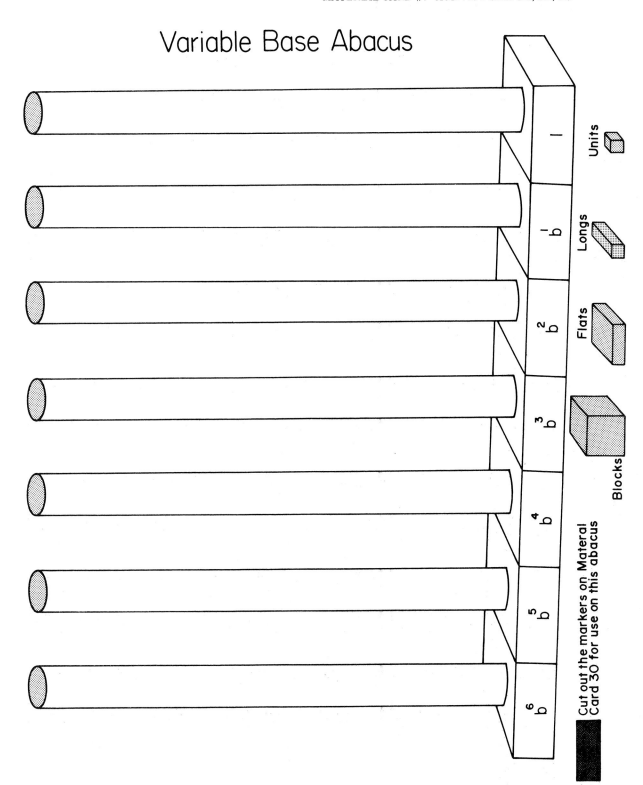

MATERIAL CARD #8—ACTIVITY SET 2.3

Mind Reading Cards

Card 4	Card 2	Card 1
4, 23	2, 23	1, 23
5, 28	3, 26	3, 25
6, 29	6, 27	5, 27
7, 30	7, 30	7, 29
12, 31	10, 31	9, 31
13	11	11
14	14	13
15	15	15
20	18	17
21	19	19
22	22	21

Card 32	Card 16	Card 8
32	16, 26	8, 27
	17, 27	9, 28
	18, 28	10, 29
	19, 29	11, 30
	20, 30	12, 31
	21, 31	13
	22	14
	23	15
	24	24
	25	25
		26

MATERIAL CARD #9—ACTIVITY SET 8.2

Rectangular Coordinate System

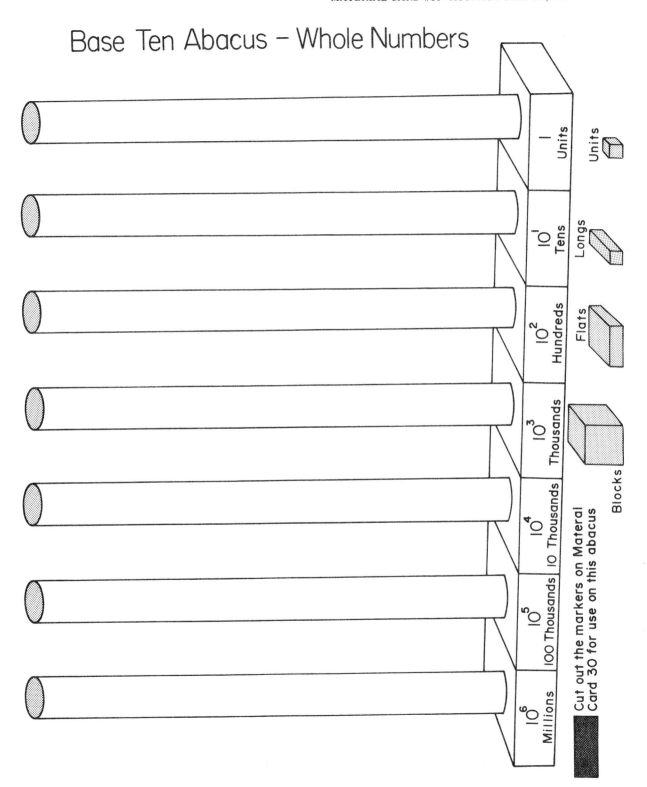

MATERIAL CARD #14—3.5 JUST FOR FUN

Rectangular Grid

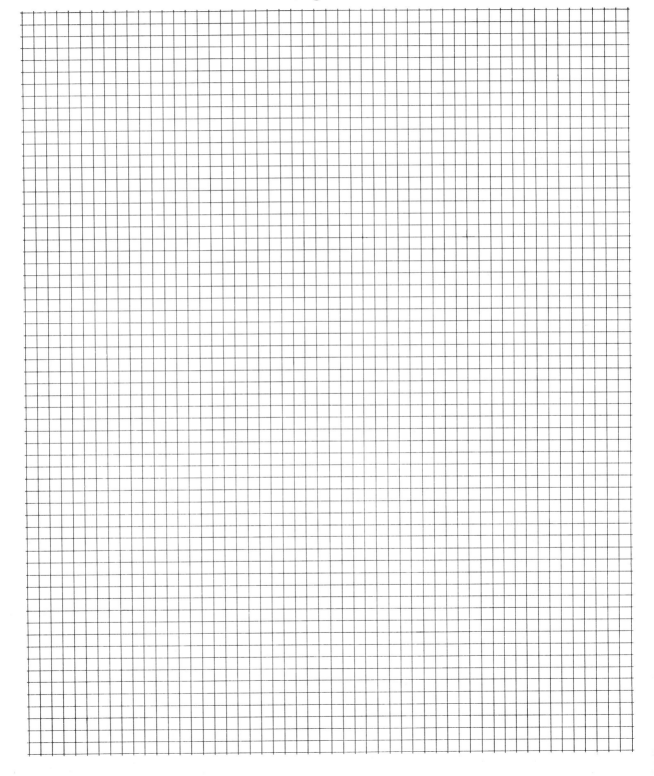

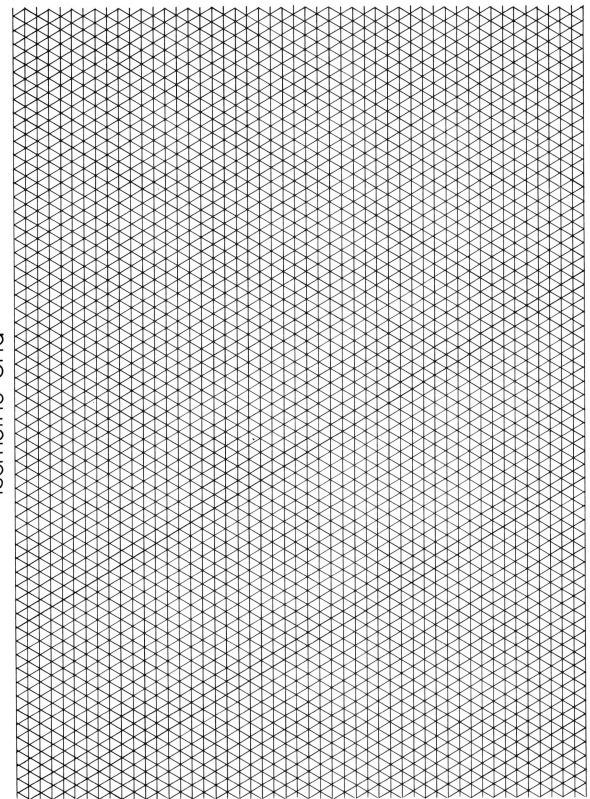

MATERIAL CARD #16—ACTIVITY SET 10.1

Three-Penny Grid

MATERIAL CARD #17—ACTIVITY SETS 4.1, 5.2, 7.4

Rectangular Geoboard Template

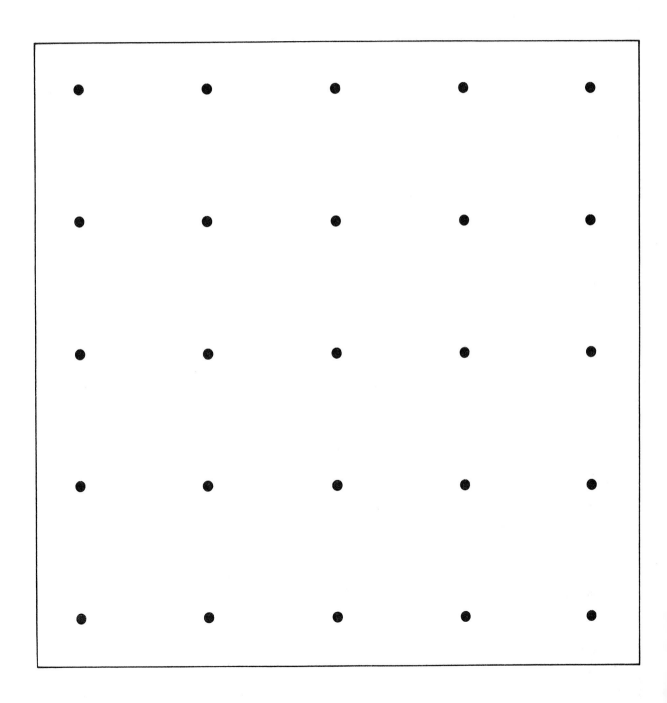

MATERIAL CARD #18—ACTIVITY SETS 4.1, 9.1

Circular Geoboard Template

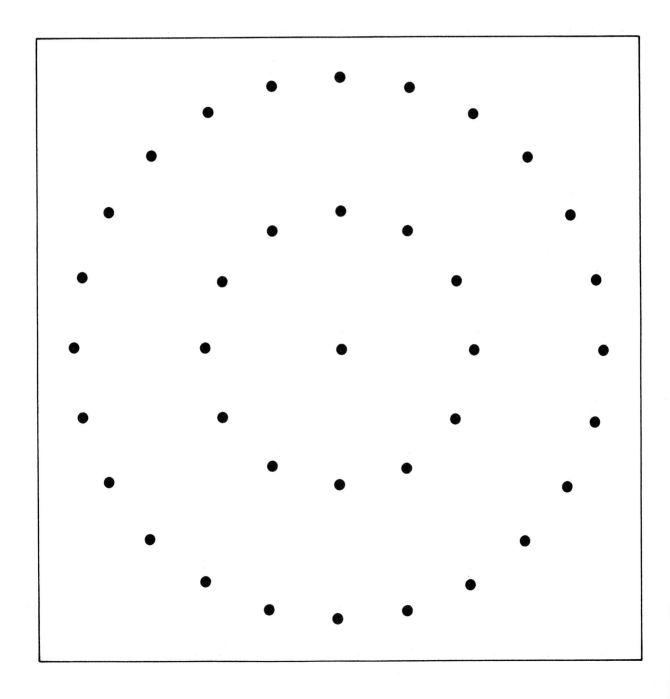

MATERIAL CARD #19−10.1 JUST FOR FUN

Buffon's Needle Problem

MATERIAL CARD #20—5.1 JUST FOR FUN

Centimeter Racing Mat

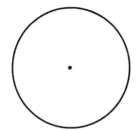
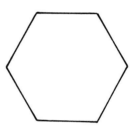
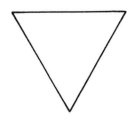
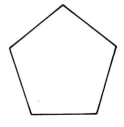

MATERIAL CARD #21—ACTIVITY SET 9.2

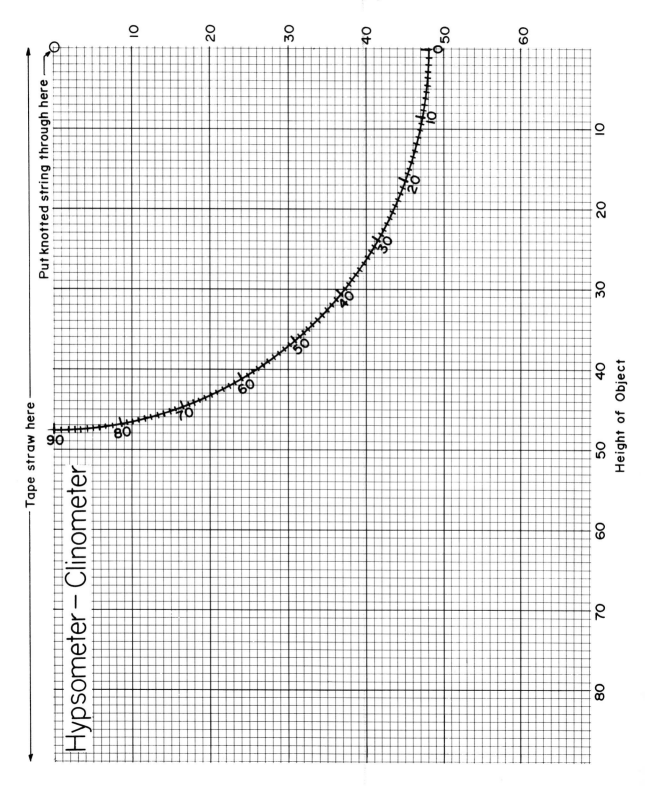

MATERIAL CARD #22–8.1 JUST FOR FUN

Algebraic Skill Cards

Team A · Team B

I have $n + 1$ **Who has** two less than a number	**I have** $4n + 5$ **Who has** two more than the square root of a number	**I have** $n + 2$ **Who has** five less than a number	**I have** $5n + 3$ **Who has** one more than three times the square of a number
I have $n - 2$ **Who has** one more than two times a number	**I have** $\sqrt{n} + 2$ **Who has** five less than eight times a number	**I have** $n - 5$ **Who has** two more than three times a number	**I have** $3n^2 + 1$ **Who has** two more than one-third of a number
I have $2n + 1$ **Who has** five less than seven times a number	**I have** $8n - 5$ **Who has** one more than four times the square of a number	**I have** $3n + 2$ **Who has** five less than six times a number	**I have** $\dfrac{n}{3} + 2$ **Who has** six less than the cube of a number
I have $7n - 5$ **Who has** seven more than the square of a number	**I have** $4n^2 + 1$ **Who has** four more than one-third of a number	**I have** $6n - 5$ **Who has** eight more than the square of a number	**I have** $n^3 - 6$ **Who has** nine more than four times a number
I have $n^2 + 7$ **Who has** three more than half of a number	**I have** $\dfrac{n}{3} + 4$ **Who has** seven less than the cube of a number	**I have** $n^2 + 8$ **Who has** five less than half of a number	**I have** $4n + 9$ **Who has** seven less than half of a number number
I have $\dfrac{n}{2} + 3$ **Who has** five more than four times a number	**I have** $n^3 - 7$ **Who has** one more than a number	**I have** $\dfrac{n}{2} - 5$ **Who has** three more than five times a number	**I have** $\dfrac{n}{2} - 7$ **Who has** two more than a number

MATERIAL CARD #23—ACTIVITY SET 4.2

Polygons for Semiregular Tessellations

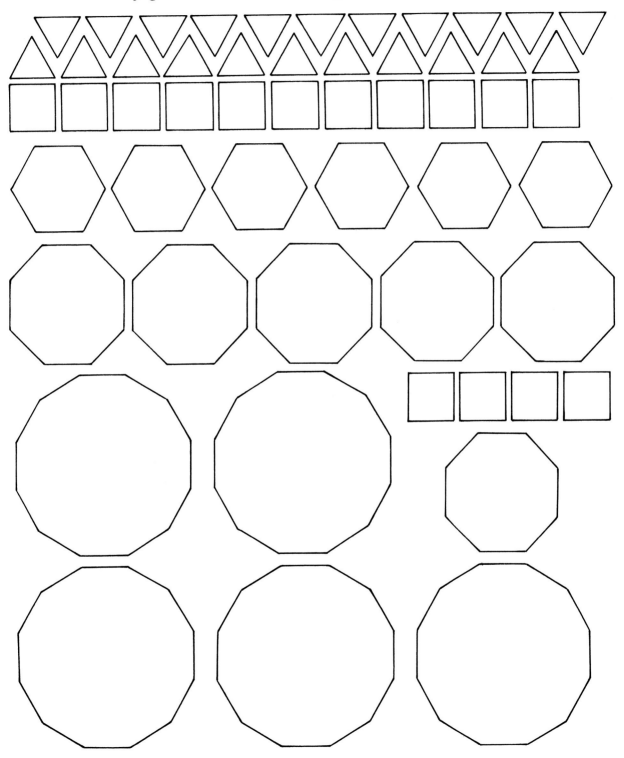

MATERIAL CARD #24—ACTIVITY SET 4.3

Semiregular Polyhedral Faces
(Use these polygons to make more faces)

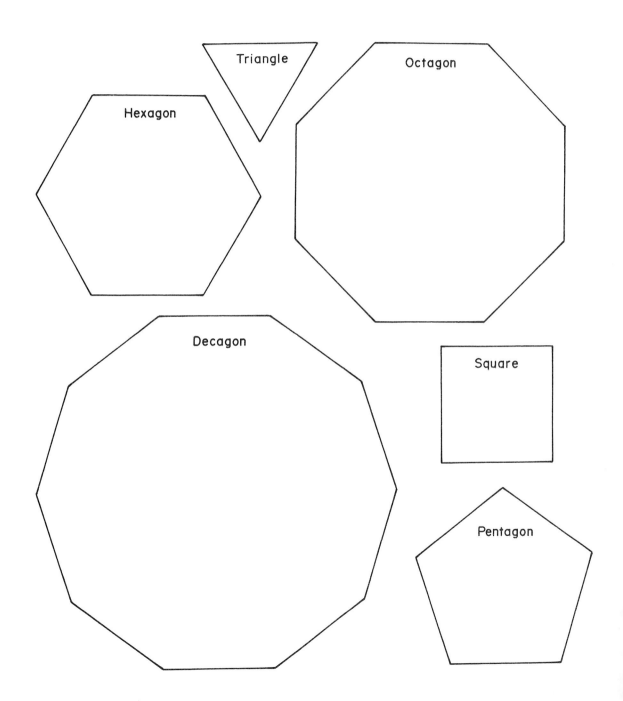

MATERIAL CARD #25—ACTIVITY SET 2.1

Attribute Pieces

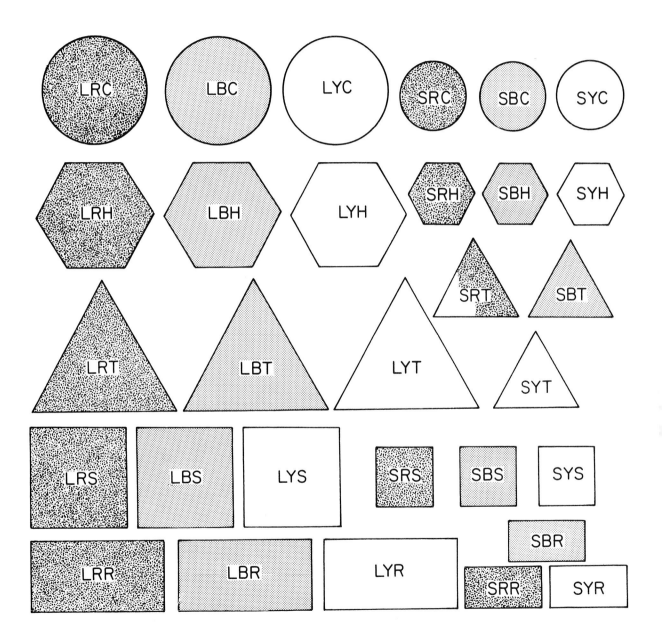

MATERIAL CARD #26—ACTIVITY SET 5.3

Prism, Pyramid, and Cylinder

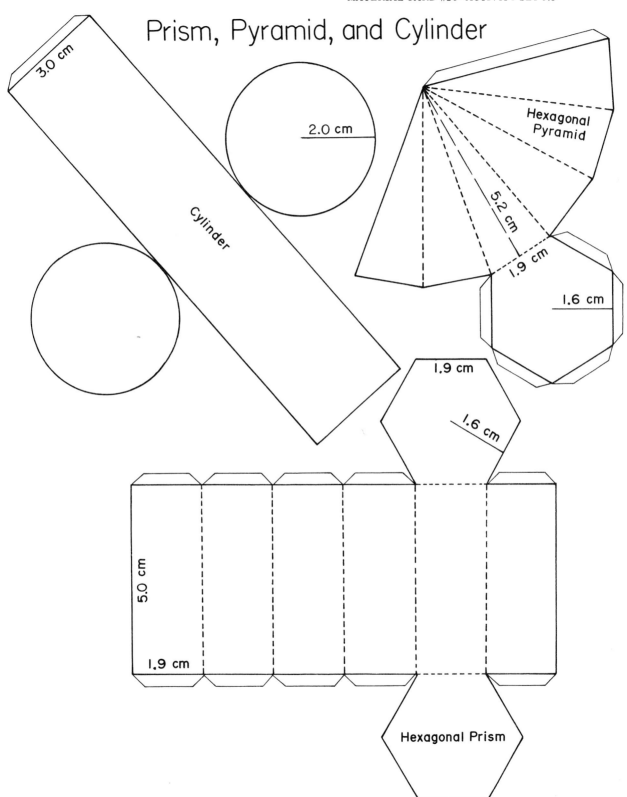

MATERIAL CARD #27–ACTIVITY SETS 6.1, 6.2

Fraction Bars

MATERIAL CARD #28—ACTIVITY SETS 6.1, 6.2

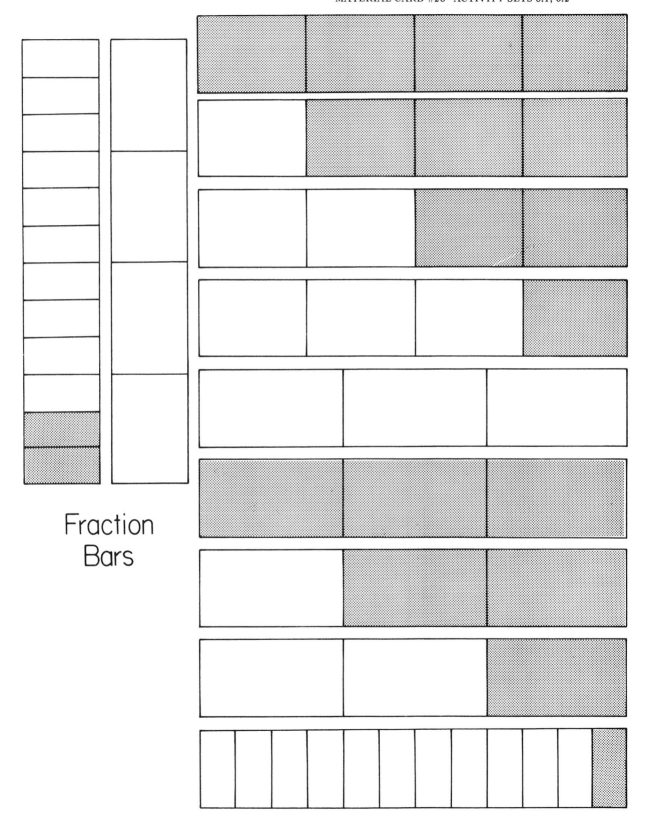

Fraction Bars

MATERIAL CARD #29—ACTIVITY SETS 6.1, 6.2

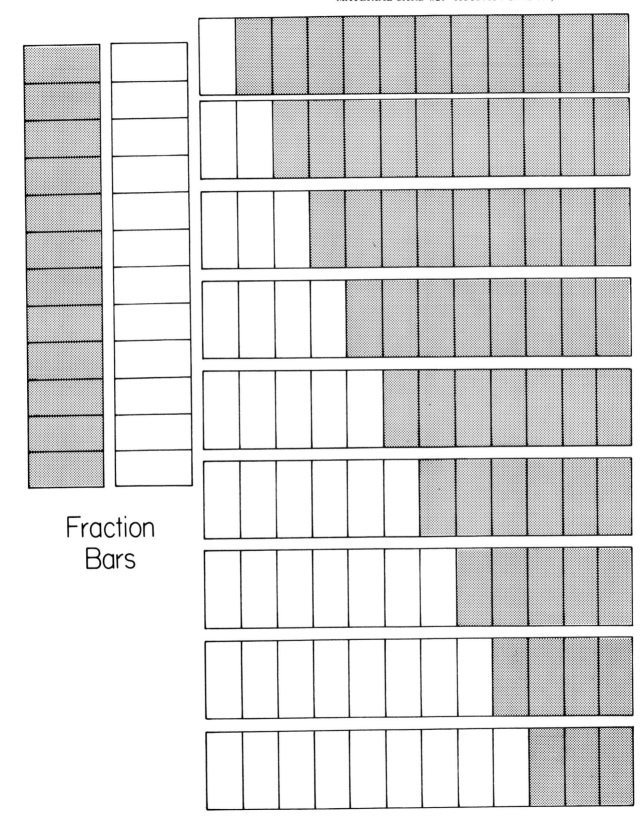

Fraction Bars

MATERIAL CARD #30—CHIPS: ACTIVITY SET 6.3
MARKERS: ACTIVITY SETS 2.2, 3.3, 3.4, 7.1, 7.2

Black and Red Chips
(Positive and Negative Integers)

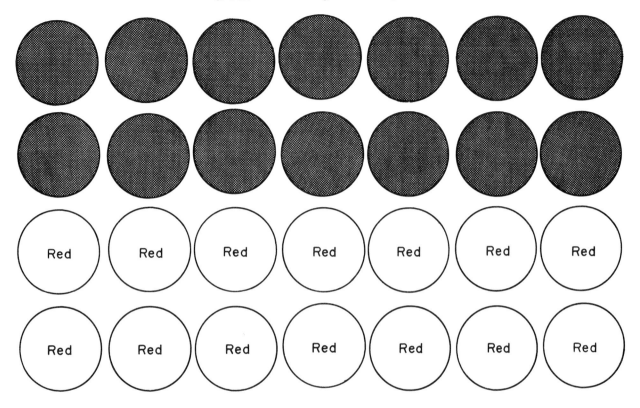

Markers for Abacuses
(Material Cards 7, 13 and 22)

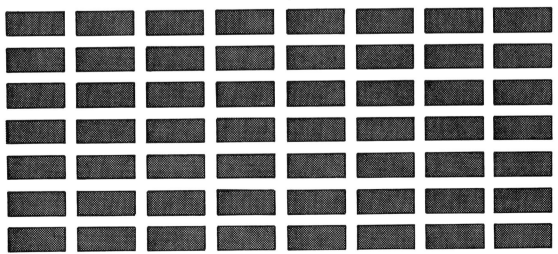

MATERIAL CARD #31—ACTIVITY SETS 2.2, 3.1, 3.2

Base Five Multibase Pieces

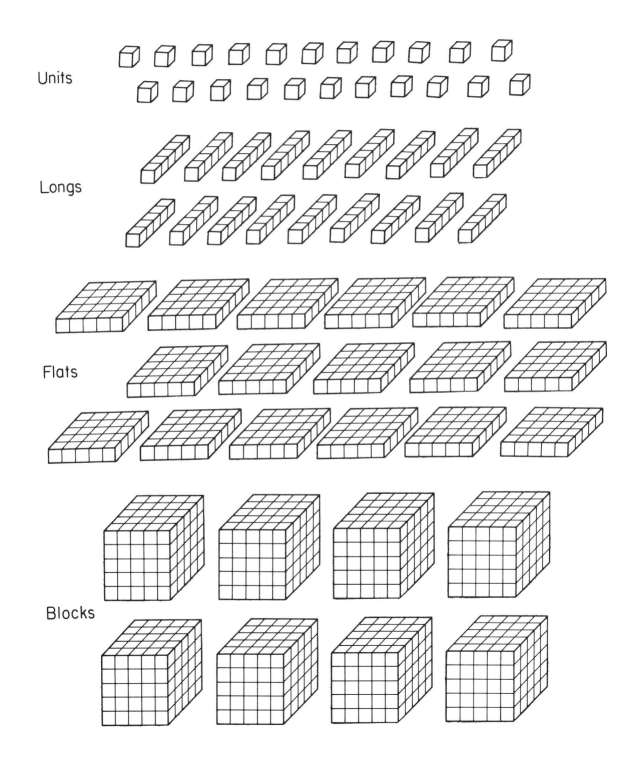

Decimal Squares

MATERIAL CARD #32 – ACTIVITY SET 7.2 and
JUST FOR FUN 7.1 and 7.2

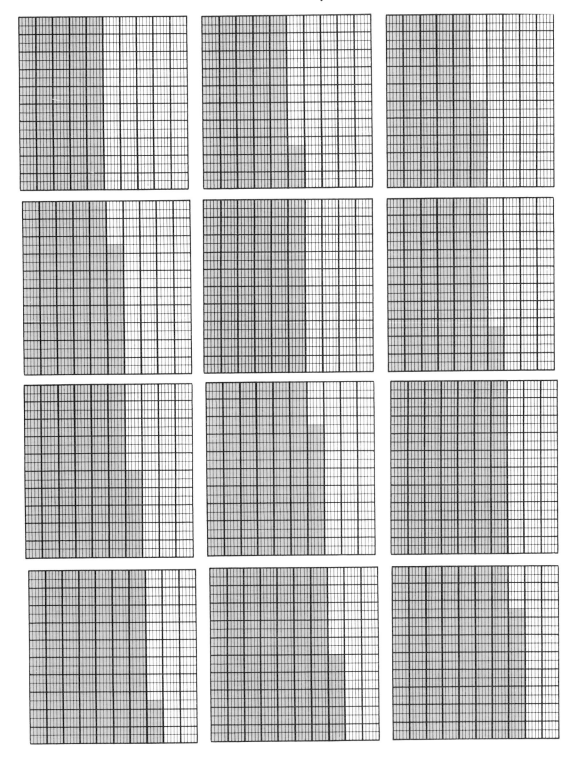

MATERIAL CARD #33 – ACTIVITY SET 7.2 and
JUST FOR FUN 7.1 and 7.2

Decimal Squares

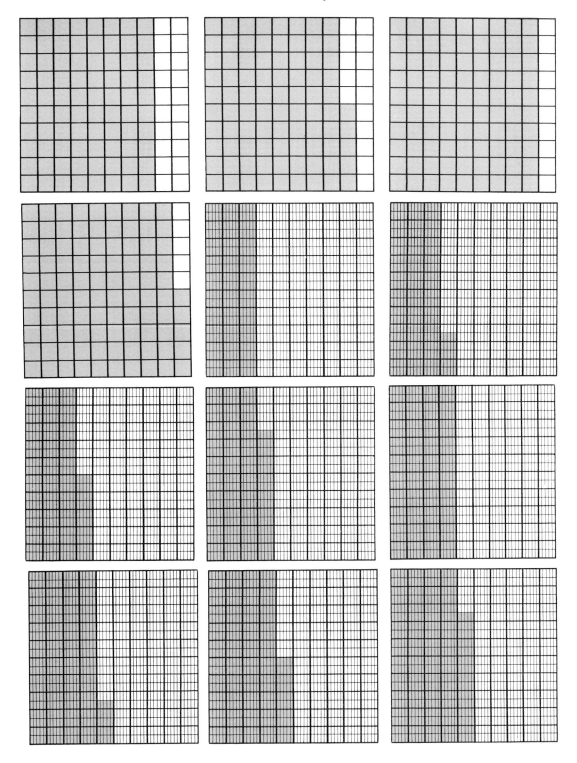

Decimal Squares

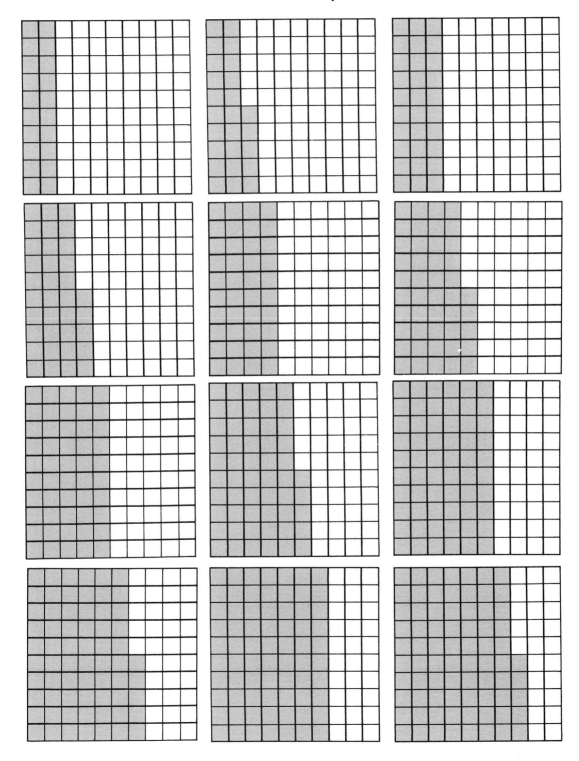

MATERIAL CARD #34 — ACTIVITY SET 7.2 and JUST FOR FUN 7.1 and 7.2

MATERIAL CARD #35 – ACTIVITY SET 7.2 and
JUST FOR FUN 7.1 and 7.2

Decimal Squares

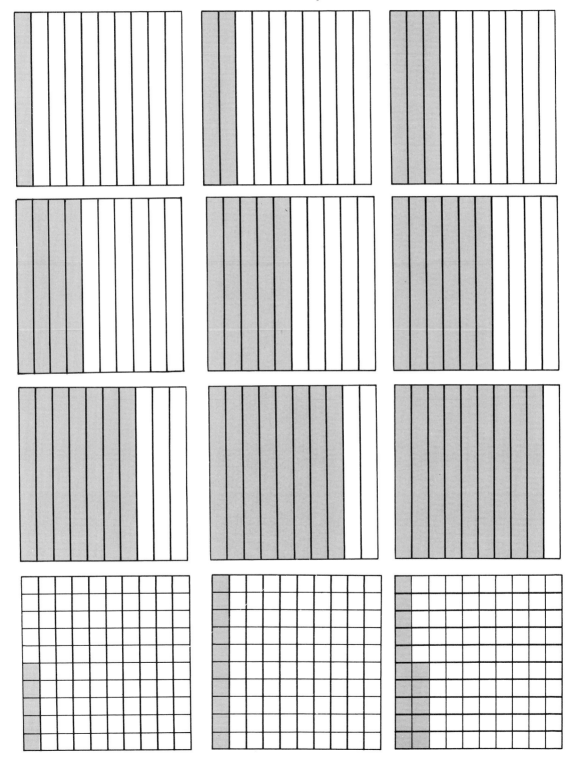

MATERIAL CARD #36 — ACTIVITY SETS 3.1, 3.3

Chip Trading Chips

yellow	yellow	yellow	yellow	yellow	yellow	yellow
yellow	yellow	yellow	yellow	yellow	yellow	yellow
blue	blue	blue	blue	blue	blue	blue
blue	blue	blue	blue	blue	blue	blue
green	green	green	green	green	green	green
green	green	green	green	green	green	green
red	red	red	red	red	red	red

MATERIAL CARD #37—ACTIVITY SETS 3.6, 6.1

Cuisenaire Rods

red	red	red	red
red	red	red	red	red	red	red	

green	green	green	green	green	

green	green		purple	yellow
green	green		purple	yellow
green	green		purple	yellow
green	green		purple	yellow

black	purple	yellow
black	purple	yellow
black	purple	yellow
black	purple	yellow

brown	red	dark green
brown	red	dark green
brown	red	dark green

| blue | | dark green |
| blue | | dark green |

| orange | dark green |
| orange | dark green |

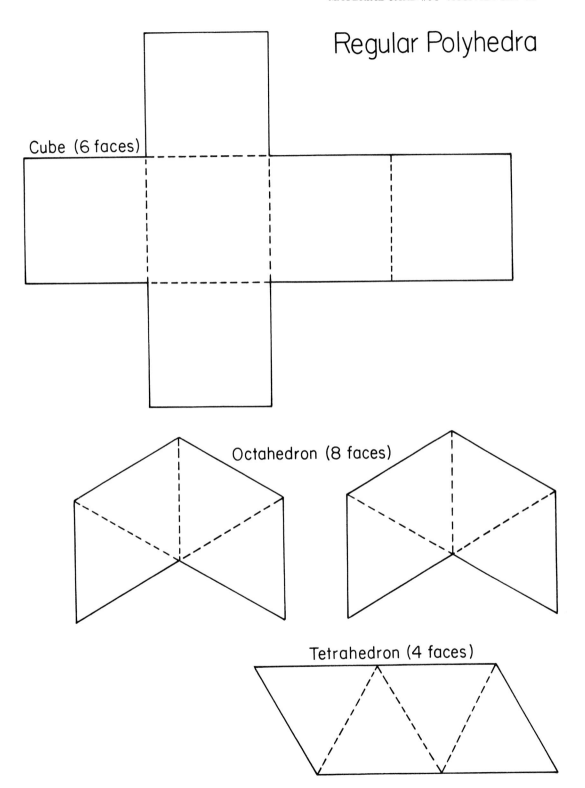

MATERIAL CARD #39 – ACTIVITY SET 4.3

Regular Polyhedra

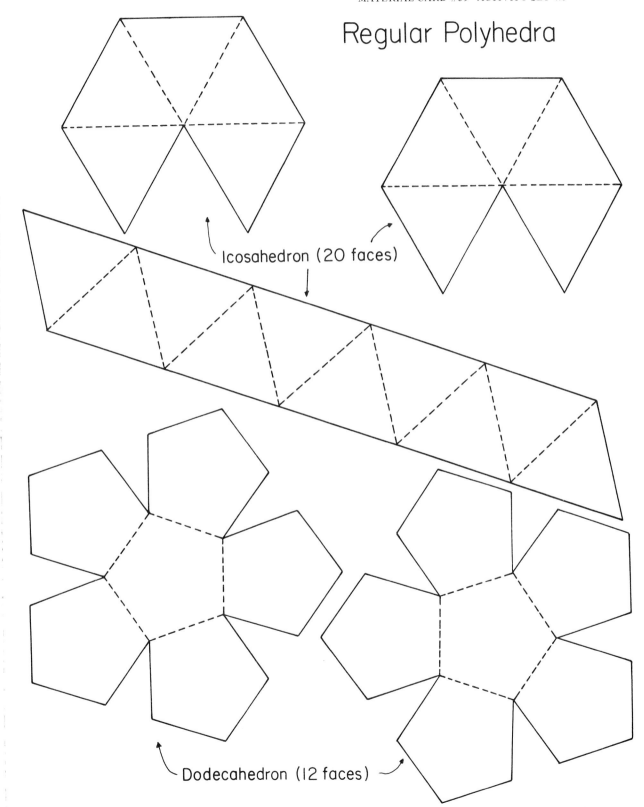

Icosahedron (20 faces)

Dodecahedron (12 faces)

MATERIAL CARD #40—ACTIVITY SET 4.3

Deltahedra—Isometric Grid

MATERIAL CARD #41 – ACTIVITY SETS 4.1, 5.1, 5.3, 8.2, 9.2

Metric Ruler, Protractor, and Compass

Compass

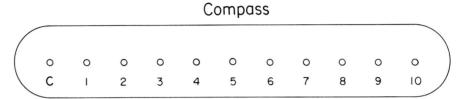

Punch out holes C, 1, 2, 3, ···, 10. Hold point C fixed for the center of a circle and place a pencil at hole 8 to draw a circle of radius 8 centimeters.

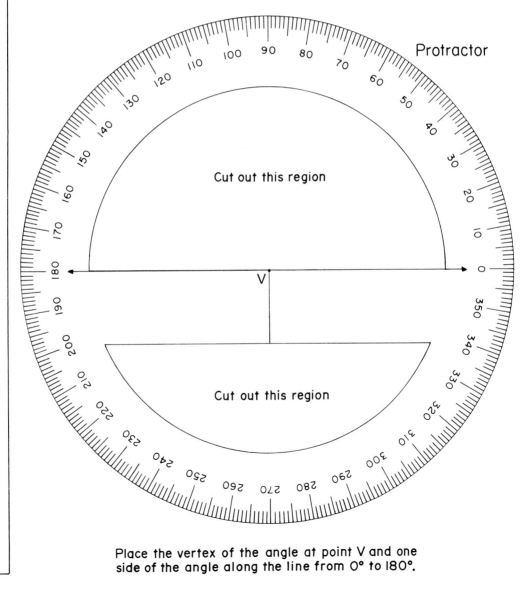

Place the vertex of the angle at point V and one side of the angle along the line from 0° to 180°.

MATERIAL CARD #42−10.2 JUST FOR FUN

Trick Dice

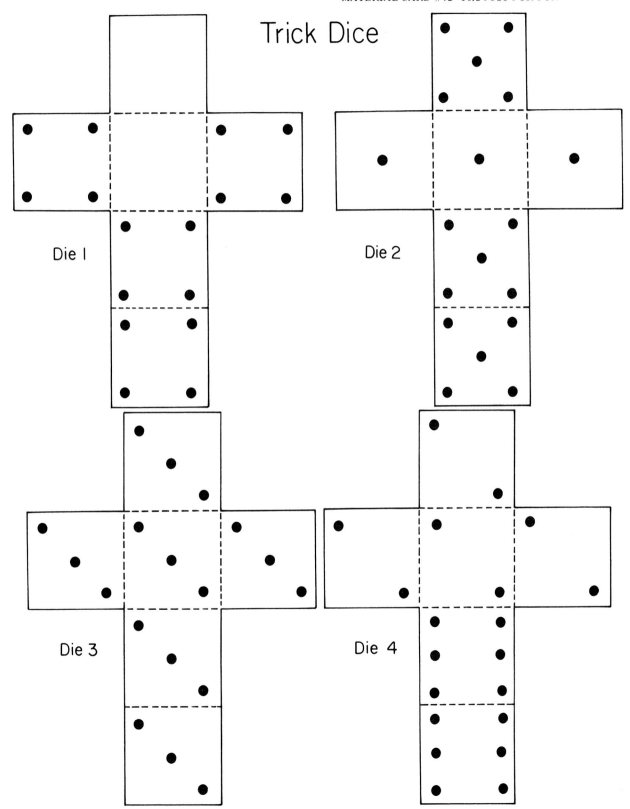

Die 1

Die 2

Die 3

Die 4